THE UNIVERSE IN TIME

THE UNIVERSE IN TIME

Paolo Maffei
translated by Mirella Giacconi

The MIT Press
Cambridge, Massachusetts, and London, England

This work was originally published as *L'universo nel tempo*,
© 1982 Arnoldo Mondadori Editore, Milan, Italy.

Translation of this book was aided by a grant from the Italian Ministry of Foreign Affairs.

This book was typeset by Achorn Graphics and was printed and bound by Halliday Lithograph in the United States of America.

Library of Congress Cataloging-in-Publication Data

Maffei, Paolo.
   [Universo nel tempo. English]
   The universe in time/Paolo Maffei; translated by Mirella Giacconi.

   p.   cm.
   Translation of: L'universo nel tempo.
   Bibliography: p.
   Includes index.
   ISBN 0-262-13236-2
   1. Cosmology.   2. Cosmogony.   I. Title.
QB981.M25313   1989   523.1—dc19   88-4113

CONTENTS

PREFACE

This book is a sequel to *Beyond the Moon*.* While I was writing that book, I realized that although it showed celestial bodies in the process of forming and stars that self-destructed and that although at some point it also became a trip back in time, a voyage in space could only portray the universe as we see it today. This would not be a big problem if everything always remained the same, or if things changed but not in their essence, as the river we see flowing by in which the passing molecules are never the same but are always water. But the universe does change, both as a whole and in its components, just as everything changes in our small world even in the short span of a human lifetime.

But how does it change? The best way to find out, it seemed to me, would be to embark on a new voyage—a voyage in time. While more difficult than the exploration of space, this venture was not impossible and in fact promised to be more adventurous and significantly richer. Exploring space and time simultaneously was not a good idea, however, because the journey would have become too long, too tiring, and above all, too complicated. Hence I began to think of a second book, and as I was writing the chapter on clusters, which by bringing up the distinction among stars of different ages made the problem more obvious than ever, I made my decision. I would postpone the voyage in time rather than ruin it by rushing through it, as occurs in certain tourist trips (Paris in three days!), and at the end of the section "Other Groups: Other Stars" I mentioned the impossibility of making both journeys at the same time.

As I wrote those lines, I felt that, rather than merely informing the reader, I was making a pledge to myself and I tried to keep it immediately by planning the new journey as soon as *Beyond the Moon* was finished. But it was not that simple. In order to understand the changes that take place in the universe and to tie them together in a reasonable chronological sequence, it was first necessary to discuss in some depth the celestial objects and phenomena that had been barely mentioned in *Beyond the Moon* (such as the origin and destruction of comets, supernovae, and anomalous galaxies) or not mentioned at all (such as the planetary nebulae, black holes, and black dwarfs). All these objects were generally strange and rare, true anomalies, discovered with sur-

*[Published in translation by D. J. K. O'Connell in 1978 by The MIT Press, Cambridge, MA.—Trans.]

prise by astronomers of different epochs or predicted by theorists but never observed. But they became very important in the new exploration because they were necessary to explain a particular phenomenon, such as the end of massive stars, or because their rarity might be due to the fact that they represented short-lived, significant stages in the evolution of the universe and its components.

For this reason, despite having started this book as far back as the spring of 1973, I had to stop and write another one. Thus was born *Monsters in the Sky,** which was meant to answer some of the questions raised by the first book and to deal with matters that, if properly addressed, would have made this third book too cumbersome and sent the reader hither and yon from the principal lines of the argument.

Like *Beyond the Moon,* these are voyages of exploration, but in time rather than in space. On the basis of concrete facts and well-established theories, such as that of stellar evolution, we shall follow the life cycle of a star, attend the birth of the sun and its planetary system, and follow the evolution of the earth and all that developed on it, particularly the life process we know only on our planet. We shall then explore more remote times, observe more ancient stars different from the present ones and the formation of galaxies, and go back in time to an instant after the universe was born. We shall discover many strange things, and starting from all that has occurred up till now, and that may have occurred in other places and other times, we shall even hazard voyages into the future—the near and the more distant future—trying to discover, bit by bit, the ultimate fate of man, the earth, and the universe.

Since it is a sequel to *Beyond the Moon,* this book is addressed first of all to the same readers, who will now be able to extend their vision from space to time; having broken the chains that bind us to our planet by means of observation, reasoning, and fantasy, by the same means they can now overcome the equally limiting bounds of our brief lives.

Extending oneself in time is even more exhilarating than traveling in space. It is a little like feeling immortal. Seeing how the universe evolved in the past and catching a glimpse of its possible futures is not only exciting but indispensable to man in his effort to understand his

*[Published in translation by Mirella and Riccardo Giacconi in 1980 by The MIT Press, Cambridge, MA.—Trans.]

place in the cosmos. Through the progressive discovery of the changes that occur in the universe, even in its more limited but better-known components such as the earth, we obtain a temporal perspective in which man appears to be more important than when he was located only in space. From all that science, and not only astronomy, has taught us so far, man emerges as a culminating point, a means by which the universe knows itself, a stepping stone for future development. All this cannot be of interest only to people who are curious about the sky. On the contrary, it is ever more evident that to understand man's place in the world, his limits, and his importance, we must first understand the universe.

The ancient Greeks chose man as a starting point for the knowledge of the universe, so much so that on the temple of Apollo at Delphi they inscribed *Know thyself*. Today we have reached the time in which we may affirm, *Know the universe and you will know yourself.*

ACKNOWLEDGMENTS

I am very grateful to all the colleagues who have contributed to this book through discussions and advice and by providing illustrative material. In particular, I wish to thank Pierluigi Ambrosetti, Alfonso Cavaliere, Domenico Faraone, Marcello La Greca, Angioletta Coradini, Italo Mazzitelli, and Roberto Nesci, who have read the manuscript wholly or in part. While thanking them for their interest and advice, I wish to point out that I am solely responsible for the scientific content of the final version.

THE UNIVERSE IN TIME

1 FROM A VOYAGE IN SPACE TO A VOYAGE IN TIME

FROM THE GALAXIES TO THE EARTH

Our trip begins on a spaceship adrift in an empty corner of the universe. Looking out of a porthole with an eye more powerful than our own, we see the darkness of space dotted with a myriad of luminous points. Since we are in a deserted region and are looking far into the distance, those specks of light are galaxies. The sight before us cannot be compared with a starry sky on the earth because the number of galaxies is at least a thousand times greater than the number of stars we can see on even a moonless night. If the galaxies were uniformly distributed in space, we would see something like a fog, but since they are grouped in clusters and superclusters (clusters of clusters) we see a sort of network, formed by poorer regions and richer ones. Figure 1 is but a pale image of our view; this map was prepared by C. D. Shane and C. A. Wirtanen of Lick Observatory by recording all the galaxies brighter than the 19th apparent magnitude.

The teeming of lights we see from our porthole recalls a little the Milky Way observed through a powerful telescope. But in that case each point is a star, that is, a body more or less like the sun; here, instead, each speck of light is at least a galaxy, a system of millions or even billions of stars.

The galaxies are not all alike (figures 2 and 3); all of them, however, are like autonomous cities separated by immense empty spaces that keep getting wider. As observations have revealed, the universe as a whole is expanding, and all the galaxies are moving away from one another in a sort of mutual flight.

Let us choose one of those luminous points and move toward it. As we draw closer, it turns into a disk, first no bigger than a pinhead, and then gradually larger and richer in detail. Finally it emerges as an enormous spiral with its arms full of stars, some scattered and some in groups, bright nebulae, and dark clouds of dust. All around this celestial flywheel are hundreds of nebulous dots, and as we pass by one of them on our way in, we perceive that it is a ball of stars, an aggregate of hundreds of thousands of stars closely packed together. This is a globular cluster, so named because of its appearance (figure 4). Almost every galaxy has a similar court of clusters, whose extent is generally proportionate to the size of the galaxy itself; some small galaxies have but a handful of them, while others, like the giant M 87, have a few thousands. Star clusters can also be seen in the spiral arms, but they differ

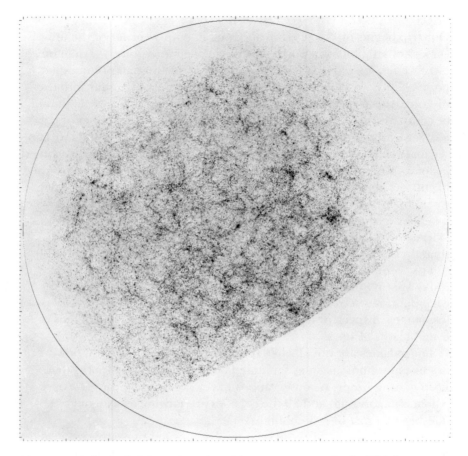

Figure 1   Galaxies brighter than the 19th apparent magnitude. This impressive map was made by dividing the sky into a large number of small squares and assigning each square an intensity proportional to the number of galaxies in it. If we could observe the universe from a region devoid of galaxies, it might look like this network of galaxies and clusters of galaxies. [From M. Seldner, B. Siebers, E. J. Groth, and P. J. E. Peebles, *Astrophys. J.* 82 (1977):249, © The University of Chicago Press]

Figure 2   Different types of galaxies in the constellation Pavo. This is how the network of galaxies in figure 1 is resolved when we move closer to one of its groups. [Courtesy the Royal Observatory, Edinburgh]

Figure 3   The galaxy M 51 in the constellation of the Hunting Dogs is one of the best-known spirals. It is also famous for having a smaller companion, with which it is interacting. [Courtesy James Wray, McDonald Observatory, University of Texas]

from the others both in appearance and in the number of stars they contain. These systems consist of a few hundred or, at most, a few thousand stars and are so spread out that they have been called open clusters (figure 5). Contained within a thin spiral-shaped layer about the galaxy's equatorial plane is an enormous amount of neutral hydrogen. Here and there, hydrogen clouds excited by hot stars glow blue, green, and, above all, red. These clouds also contain other elements, such as oxygen, helium, nitrogen, and sulfur, and sometimes even molecules. Often, such gaseous nebulae are associated with dust and groups of hot stars or with aggregates of stars of similar characteristics, known as associations (figure 6).

   Although our attention is caught by all these large and bright features of the galaxy, we should not overlook the smaller bodies, not easy to see, but of greater interest to us because they might harbor life—

Figure 4   The globular cluster 47 Tucanae, 13,000 light-years away, photographed with the 4-meter telescope of Cerro Tololo. It is visible with the unaided eye. [Courtesy V. Blanco, Cerro Tololo Interamerican Observatory]

namely, the planets. While we know with certainty of only nine planets (those of the solar system), there is both direct and indirect evidence of the existence of planets belonging to other stars. If only half, or even fewer, of all the stars in the universe had planets, their total number would be enormous because there are millions of galaxies, each containing billions of stars.

A galaxy is not a static system. All its components are in motion, and as a whole it rotates about an axis that passes through the galactic center at right angles to the equatorial plane. Nor is it immutable. Its appearance gradually changes, partly because its components are in motion and partly because they undergo transformations with the passage of time. Some change so slowly as to make any difference imperceptible over a short period of time. But there are also pulsating stars, which vary in a periodic fashion, and novae, which seem to spring suddenly from nowhere (though they are in fact existing stars that are

Figure 5   The open cluster M 16 discovered in 1746. The photograph shows that it is embedded in a nebulosity where bright regions are mixed with dark dust clouds. [© 1982 California Institute of Technology]

Figure 6　The association at the center of the nebula M 42. Some of the stars are newly born; others are still surrounded by dust cocoons visible in the infrared. [Courtesy Lick Observatory]

ejecting their outer layers into space). And then there are stars that undergo catastrophic changes—the supernovae, which self-destruct in a gigantic explosion.

Galaxies differ in size, structure, and composition, but we shall explore only one, the galaxy we have just entered. Let us suppose it is our own Galaxy. It is a fairly typical spiral and one of the largest, with a diameter of 100,000 light-years and a maximum thickness of 16,000 light-years.[1] In this vast expanse, in which empty space far exceeds the space occupied by matter, we must try to locate the sun. To do so, we shall navigate to a spot in a spiral arm 30,000 light-years from the galactic center. It will not be easy to find it, but if we succeed we can reach the earth by making directly for the center of the planetary system to which it belongs. Our planet is so close to the center, in fact, that even from Pluto's orbit it does not seem to be that far from the sun, which at this distance we see only as a very bright dot. After crossing the orbits of the outer planets we arrive in sight of the earth and keep approaching it until our motion through space becomes a descent to the body below us, which fills our entire field of view and looks immense, just as intergalactic space with its web of galaxies looked a little while ago. At a certain point we touch down on its soil; all around us we see a horizon and, above it, the sky, in which is displayed the half of the universe that is above our horizon.

Our long voyage in space is ended. We are back home, and from here we shall begin a new and even more exciting voyage—a voyage in time. Without leaving our native soil, we shall explore the past and glimpse the future of the earth, the sun, the stars, the galaxies, and finally the universe as a whole. These bodies will pass before us, evolving in front of our eyes, though our view will unfortunately be neither complete nor definitive but will come only in flashes. But these flashes will amaze us, particularly considering that we have managed to produce them from our limited and isolated earthly observatory in a period of observation that is minuscule compared with the immense intervals of time that characterize the vast transformations of significant parts of the whole universe.

OF TIME AND CYCLES

The existence of time is an intuitive truth. Our actions unfold in time, and this very fact makes it as palpable for us as the space in which we

move. As in the case of space, however, we can define time only operationally, that is, by measuring it.

From any point on the surface of the earth, if the day is not cloudy,
we see the sun. It will not take us long to realize that the sun is not
stationary. Slowly, it moves across the sky in an arc of circle and then
dips down toward the horizon and sets. If we wait a while, we shall see
the light blue sky turn dark and the stars appear and move along in the
same direction as the sun. They, too, rise, travel in more or less wide
arcs of circle, and then set. In time the sky begins to lighten in the east,
at first imperceptibly, and then enough to erase the stars, and the sun
rises again above the horizon, flooding the earth and the sky with light.
After traveling a similar path across the sky it sets again. At this point
we shall have the feeling that a cycle has been completed, that is, a
series of events that differ from one another but repeat themselves in
the same way and in the same order. And indeed it is so. After the sun
has set, the night will return with its mantle of stars; the sun will rise
again, travel another diurnal arc, and set; and so on. By counting a
number of similar cycles we can measure the interval of time in which
the cycles repeat. We can use intervals of any length; we just have to
express them with larger or smaller numbers.

The daily cycle must have been the first discovered by man. Actually,
one cannot say that man "discovered" it because he was regulating his
life on this cycle long before rationalizing it, even before standing erect,
for the simple reason that light and darkness determined his periods of
activity and rest just as they did for all the other animals.

There are longer cycles that can be easily discovered and that furnish
larger units for the measurement of time. If we watch the sunset every
evening from the same place, we shall readily notice that it does not
always occur at the same point (say, that small faraway house on the
hill that barely rises above the horizon); every day it shifts a little more,
for example, to the north.[2] During the same interval the point where
the sun rises also shifts to the north, so that each day the sun travels a
wider arc, climbs higher in the sky, and remains longer above the horizon. As a result, little by little the days grow longer at the expense of
the nights, which become shorter. The day arrives, however, when the
points on the horizon where the sun rises and sets stop advancing, and
soon afterward they begin to move back, toward the south. As they
shift southward the path of the sun across the sky becomes increasingly

shorter, and consequently the days grow shorter. Eventually the diurnal shift toward the south stops, only to start again in the opposite direction, and one day the sun sets at the same exact point where it had set on the evening of our first observation, behind the small house on the faraway hill. Of course there was another evening, during the return journey from north to south, when the sun set in the same place. But only now that the sun is back at the starting point after completing the entire excusion along the horizon can we say that the cycle is complete.

As this wider cycle unfolds, certain phenomena occur that are as evident as the alternating of day and night and that certainly did not escape the attention of primitive people. The amount of heat that reaches a point on the earth's surface in the course of a day is greater when the sun remains long above the horizon, while the same point receives less heat when the sun's daily journey is short. Consequently, if we remain in the same place we shall see cold days alternate with warm days, with mild but unsettled days in between. The cave dweller did not need astronomical observations to understand this seasonal cycle any more than does modern man confined to his house or place of work.

Time also may be measured by means of a third cycle of intermediate length, equally evident. Even primitive men must have noticed that the end of the day is not always followed by total darkness. Some evenings, as the sun sets, a luminous disk rises in the opposite direction, not as blinding as the sun and even marred by dark spots, but bright enough to bathe the landscape in a silvery light. This disk (which looks roughly half a million times less bright than the sun) is not always present in the sky, does not always have the same shape and brightness, and does not always rise as the sun sets. On the next evening, in fact, it rises later and is not perfectly round. On each successive night it rises still later and looks less full, until it becomes a thin sickle that can be seen only in the light of dawn, just before sunrise. Then it disappears. A few days later it reappears in the evening sky still in the shape of a thin sickle but with its luminous side reversed, so that it always faces the sun. On each subsequent evening the sickle appears fuller and farther away from the sun, until finally it becomes a full disk again, which rises as the sun sets. The cycle is now completed and begins anew. I do not have to tell you that this disk is the moon and that we have been observing the lunar phases and the lunation.

These three cycles are the sources of our day, year, and month, and together they form the frame of reference for all our activities that we call the calendar. Unfortunately, the year does not consist of a whole number of days or a whole number of lunar months. This has led to the establishment of different calendars, such as our current calendar based on the sun alone, that of the Moslems based only on the moon, and the Hebrews', which takes both into account. Our calendar is based on the sun, which determines the day and the year, and the introduction of the leap year has taken care of the fraction of a day left over every year. The lunar month, however, has fallen into disuse almost everywhere, and in its place we have months of 30 or 31 days that have nothing to do with the lunar cycle, which repeats throughout the months without any fixed coincidence in the calendar dates (since they have different lengths). Therefore it does not make any sense to speak of a January or a February moon, and so on, as they still do in the Italian countryside in an effort to reconcile two incommensurable cycles such as the solar and the lunar.

The idea of cycles was accepted readily from the earliest times, not only because of practical or astronomical considerations, but also for cultural reasons. Man has always had a great respect, almost a cult, for all that is cyclical because the idea of cycle removes the need for conceiving an absolute beginning and end. With a cycle, everything ends but then everything begins again. In the past, as well as today, the unfolding of the celestial phenomena we have examined tended to reinforce this conviction, because the changes one observed in the sky (that is, in the cosmos) were seen to be cyclical. If a phenomenon did not seem to follow the rule, it was interpreted as belonging to a cycle so long that at the moment it could not be perceived. Throughout man's history this conclusion has been reached many times.

Furthermore, because of this psychological need, reinforced and sanctioned by the discovery of celestial cycles, man also has adopted nonastronomical cycles, such as the week, and even introduced some that have no connection whatever with practical life.

In sum, the astronomical cycles fulfilled one of man's fundamental needs. And even today, through the alternation of work and recreation, in our weekly and annual holidays, and in many other ways, we constantly try to exorcise the passing of time by hanging on to our cycles, which gives us the feeling that everything will return and that tomorrow or next Sunday we shall have what we did not succeed in having yesterday or last Sunday.

All that is an illusion. The sense of the absolute and the eternal that we read in the celestial cycles, and is reflected here on the earth in the alternation of days and nights, lunar phases, and seasons, persists only as long as our knowledge of the universe is limited. Things are entirely different as soon as we trace the cycles back to the basic phenomena that produce them.

Day is a consequence of the rotation of the earth about its polar axis; the lunar month is due to the revolution of the moon around the earth; the seasons and the year are due to the inclination of the earth's axis and the revolution of the earth around the sun. If we just move to the equator or one of the poles the phenomena change, though remaining cyclical in nature. But if we leave the earth and move to any point in space, everything vanishes: there are no longer days, or lunar months, or years, or seasons. The cyclical phenomena we have observed are a consequence only of our particular position on the surface of the planet earth and are not at all universal.

Actually this is not immediately evident. Let us look at the solar system from a point in space located on the perpendicular to the earth's orbital plane, at some distance from the sun and stationary in relation to it. By observing the sun, earth, and moon we can easily find our diurnal, lunar, and annual cycles: We only need to watch the earth rotate about itself, the moon circle the earth, and the motion of both around the sun. Indeed, it will seem to us that these cycles no longer can escape us because we have caught them in their essence—the regular and periodic motions of bodies in space. With suitable measuring devices we find that our planet completes a full rotation in the interval that on the earth we call a day; after a certain number of days the moon will have completed a full revolution around the earth; after a longer period, but still equal to our year, the earth will have completed a turn around the sun and will retrace its steps, over and over again, as it has done since time immemorial.

Not true. The rotational velocity of the earth varies continually over the years (figure 7a) and in the course of a year (figure 7b) and sometimes abruptly and suddenly. There are various reasons for this and not all very clear—ocean tides, seasonal movements of large air masses,

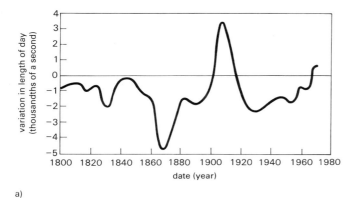

a)

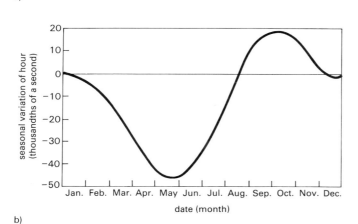

b)

Figure 7   Variation in the duration of a day: (a) in the course of nearly two centuries; (b) in the course of the seasons of one full year. [From *Le Scienze* (March 1972)]

seasonal changes in vegetation and snow cover, and even strong earth-quakes. Since the earth does not remain the same from one day to the next, there are no two days of exactly the same length. Naturally these variations are very small—a few thousandths of a second—but they are there, and over thousands or millions of years they could have critical consequences. For example, they could slow the earth's rotation to the point that a day would be as long as a year. In that case the earth would always turn the same side to the sun, and on that side it would always be day; on the opposite side it would always be night, and the diurnal cycle would no longer exist.

In the other cycles, too, what has occurred never repeats itself exactly. Contrary to what our school books show in order to illustrate the lunar phases, the moon never closes its circle around the earth. As it travels around the earth in the course of the lunation, the earth moves on its orbit around the sun and the moon is forced to accompany it on an orbit that for this very reason can never be closed. Furthermore, the moon's orbit, assumed to be closed, is not a circle but an ellipse, inclined to the plane of the orbit that the earth travels around the sun, and its characteristics, such as eccentricity, the points of maximum and minimum distance from the earth, and inclination, vary continually. All these variations combine to produce other cycles, for example, the cycle of the solar and lunar eclipses. But in the long run these cycles, too, are seen to change, form, and dissolve.

The same can be said for the yearly cycle. In the first place, in the earth's orbital elements there are variations of the type mentioned for the lunar orbit. Furthermore, not even the earth succeeds in annually closing its circuit around the sun. This is because the sun also moves; it revolves around the center of the Galaxy at the speed of 250 km/sec (kilometers per second), completing a full revolution in 226 million years. To follow the sun, the earth must keep moving onward in space, and therefore it never retraces its steps as it appears to do when we draw the closed elliptical orbit that assumes the sun as stationary. The lunar months and the years always seem to be perfectly identical on the sky, but in reality neither the moon nor the earth ever passes twice through the same point in space—not even after a full revolution about the galactic center, since the path of our sun and its system also does not exactly repeat itself. The reason is that the relative positions of all the stars and the other galactic objects change, the distribution of the

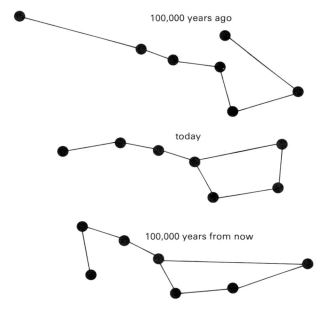

Figure 8   The constellation Ursa Major in three different epochs.

masses within our galaxy changes, and consequently, little by little, the orbits of all its members change.

This last statement seems to be refuted by fact because the constellations remain unchanged, so that if we look at the sky every year at the same hour, we see the same stars in the same positions. This is also an illusion. It is only because of the stars' great distances from us that we cannot perceive even marked displacements. In effect, by measuring their positions on the celestial vault and repeating these measurements at long intervals of time it can be shown that the stars are not "fixed" as the ancients believed. Thus the constellations do change in appearance. One example will suffice. Compare the way Ursa Major looks today with the way it looked 100,000 years ago and the way it will look 100,000 years from now (figure 8).

Everything changes; nothing remains the same; nothing ends and begins again cyclically. And not only in celestial motions. The sun, our own sun that every day looks exactly the same as the day before, not only is constantly changing in its surface appearance, as telescopic observations show, but is continually dissipating part of its mass in space in an irreversible process. We think we are watching the same sun set, but instead every evening the red body sinking below the horizon weighs 370 billion tons less than the day before. That is the amount of mass that the sun has converted into energy and dissipated in 24 hours, while in its interior a large amount of hydrogen turns into helium.

What happens to the sun happens to all the stars, even if they look the same every night. In effect, most of them, including the sun, do not change in brightness from one night to the next (and later we shall learn the reason why), but every 24 hours each of them loses an enormous amount of mass, sometimes greater than the entire mass of our planet.

In conclusion, none of the cycles that first imbued man with the sense of nature's unchangeableness is in fact general, constant, and eternal. More important, none of them has the power to reduce time to a closed circuit that would allow some phenomena to repeat in the same manner.

According to this new and more realistic view, time appears as something that sweeps all things along, changing everything and repeating nothing, as a dimension that one travels in only one direction, that is, only once.

Apart from these considerations, which we shall bear in mind, we can define the day, the month, and the year in a conventional way—as well as the hour, the minute, and the second for shorter intervals—and we can use these units to measure cosmic events as we do the events of everyday life. To this end, let us imagine an ideal time, purely mathematical, marked by a clock of perfectly uniform motion that advances in precisely equal seconds. This "standard second" can easily be obtained from astronomical observations by assuming that its duration is a fraction of a given hour, day, and year. This is the method that was used in 1957 to establish the length of a second. However, in 1964 even this astronomical connection was obviated by the introduction of atomic time, which yields a standard second that is constant and always available. With this new second, the astronomer, who was once the high priest of time as the vestal virgin was of fire, can measure changes in the second obtained from the earth's day, that is, changes in the rotational velocity of the earth.

The introduction of the atomic second shows that there can be nonastronomical ways of measuring time. Furthermore, quantum theories and the theory of relativity have introduced different kinds of time. The time we have defined (and its properties) is valid for the range of large numbers by which our daily life is regulated; it also can be used for exploring the universe, provided we do not observe a few atoms at a time and provided we do not fall into a black hole, something we should definitely try to avoid.

RETRACING THE EARTH'S PAST

While celestial events mark the time with their cycles, imperfect and relative as they may be, there are many other things here on the earth that give us the feeling of the flow of time. Life itself offers us various units of measure—from the shorter ones, of the order of a second, given by the heartbeat and the intake of breath, to those of intermediate length, of the order of a day, given by the need to eat and sleep. For longer periods, of the order of a year, we need at least an effect of the astronomical cycle, such as the changing of the seasons. The passage of decades, on the other hand, again can be observed in ourselves and in our fellow beings through the changes that our bodies undergo: the spurt of growth in infancy, the disquieting changes in face, voice, and desires in adolescence, the slow maturing in the middle years, and then the first signs of old age—white hair, baldness, and wrinkles. All this takes only tens of years, a very short time even though it comprises our entire life: all the time that is given us to do and experience things that seem so numerous and so important and to glance at the universe for an instant so brief as to give us the illusion of its unchangeableness and eternity.

Successive generations mark yet longer intervals. While many individuals appear and disappear, slow but significant changes take place in man's environment, habits, and way of thinking. The records left by our ancestors and handed down through the centuries tell us of the rise and fall of other cultures and civilizations in different lands and in remote epochs. These records show that not all that exists or is known today existed or was known in the past. This is not too surprising, particularly in this day and age when everyone, however young, can recall some novelty that has been introduced in his lifetime. But as we go back in time, changes become much more impressive because there are fewer and fewer familiar things, and even those look different from the way they look today.

Starting from the present and working backward, the first things that disappear are the products of our technological civilization, but others appear because the past was no less rich in art, science, and thought. Many are the things conceived and built a few hundred or a few thousand years ago that were later replaced by other products and are now remembered only from old documents and books. All this is true, however, only for the past few thousand years. As we go back further, the environment created by man changes radically. First the large cities grow smaller and fewer in number; then, little by little, many other things change and disappear—irrigation canals, dams, cultivated fields, houses, and finally huts, domestic animals, and crops, all products of selection. All it takes is a small step backward of tens of thousands of years. In a short while man himself disappears, his kind having become indistinguishable from the other groups of primates; a few more steps backward and even the apes and protoapes are gone. Everything has changed, including the celestial vault, because the earth along with the solar system is in a different part of the Galaxy and not all the stars we see are the same stars we see today, or they are rearranged in a different way, forming new constellations.

The large-scale features of today's landscape also have changed in a radical way. There are still mountains and seas, but you will not find the green valley or rocky outcrop you see today from your bedroom window; in its place there may be a desert or a very tall mountain, and perhaps your house could not have been built where it is today. There may be no land at all because the continent on which you dwell does not exist yet or, if it exists, it is in a different part of the earth and has not yet moved to its present location. When we go back tens of millions of years we find that continents and oceans are different and displaced. As we go further back in time the land masses move closer to one another until, about 200 million years ago, they are all joined together in one vast continent called Pangaea, which means "the whole earth."

At that time life on earth was different from what it is now, and it becomes increasingly different the further back in time we go, until we reach a time in which there is no trace of life anywhere on the terrestrial globe. At that point the only things we would find similar to what we know today are rock, lava, and poisonous volcanic gases.

Records of all the ages we have quickly crossed can be found by slicing through the earth's crust and reading the great book of rock strata, which in its various levels, as in the pages of a manuscript, contains precious evidence of the past (figure 9). Imprints of leaves and branches, fossil tree trunks, animal skeletons, and many many other relics tell us what a place was like at a particular time and what lived there. Stratigraphy also can tell us how that place and its inhabitants changed in the course of time. If the sedimentary bed has not been disturbed, the most recent layers are at the top, followed by increasingly older ones as we move downward. Thus plant and animal species can be put in a chronological sequence depending on the layer in which their fossil remains are unearthed. In addition, by finding in different places the same strata, or strata containing the same fossils, we can reconstruct the appearance of the earth at the time each layer was forming. We can discover which lands were above water and which were submerged, which lands formed vast plains and which rose high as mountains or plateaus. We can even obtain information about the climate. The study of strata, in sum, tells us what our planet was like in the various geological eras and in what order they followed one another.

The work of reconstructing the history of the earth and its inhabitants began two centuries ago. But while the strata yielded a chronological sequence, they could not easily be made to reveal their ages. Although four eras were recognized, each subdivided into periods, no accurate method could be found to date rocks and fossils or to estimate the lengths of the various eras and periods and the intervals of time separating them from us. Furthermore, the age of the earth was not known, and therefore there was not even a starting point from which to measure geologic times. It is only in the last thirty years that these problems have been solved thanks to a clock that has been working with the same precision since the birth of our planet. This clock, which is not astronomical, furnishes a way of measuring time with great accuracy that is independent of the motions of celestial bodies.

As you probably know, matter is composed of atoms and each atom, according to the simplest model, consists of a positively charged nucleus and a number of peripheral electrons that carry (identical) negative electric charges. The number of peripheral electrons is such that the sum of their negative charges neutralizes the positive charge of the nucleus. The latter, in turn, consists of neutrons and protons, which

Figure 9   Geologic strata formed by sedimentation over millions of years. Like the pages of a great book, they contain the records of the earth's past. [Courtesy Roberto Colacicchi]

are heavy particles of the same mass. The neutrons, as the name implies, are electrically neutral, whereas the protons carry a positive electric charge of the same value as that of the electrons but of opposite sign. The number of protons in the atomic nucleus of an element determines its chemical and physical characteristics and is called the element's *atomic number.* Changing the atomic number means passing from one element to another. If two or more atoms of the same element have nuclei of different weight, this means that the number of neutrons in their nuclei is different. In this case they are called isotopes of the element. Thus isotopes are atoms having the same atomic number but different *mass number,* which is the sum of neutrons and protons (and is written next to the name or chemical symbol of the element).

Let us consider, for example, the atom of helium, which has two protons and two neutrons in the nucleus and two peripheral electrons. If we remove one of the electrons it becomes an atom of helium ionized once; if we remove a proton as well, it becomes an atom of hydrogen of mass greater than normal, called tritium; if we now take away a neutron it becomes an atom of heavy hydrogen, known as deuterium; and if we take away the second neutron as well, we obtain an atom composed of a proton and an electron. This is the atom of ordinary hydrogen, while the previous two are rare isotopes of hydrogen.

There are unstable elements in nature in which the mass and electric charge of the nucleus, as well as the number of peripheral electrons, change in time. In brief, these are elements that spontaneously turn into others. Let us take as an example the atom of uranium, the most complex atom found in nature, and more precisely one of its isotopes, uranium 235, which consists of a nucleus of mass number 235 and 92 peripheral electrons. This atom is unstable; that is, it gradually loses some protons, neutrons, and electrons and turns into lighter atoms of lower atomic number until it becomes lead of atomic number 82 and mass number 207. The time needed for a certain amount of uranium 235 to turn into lead 207 can be easily calculated. What has been found is that it always takes the same time to turn half a certain amount of uranium 235 into lead 207. This period of time is called half-life.

Because of this property, uranium can be used like an hourglass that measures time by the number of grains of sand that have passed from the upper to the lower part. If we put a certain amount of uranium 235 into a box and some time later we open the box, from the number of

Table 1   Principal radioactive isotopes used to determine the ages of rocks

Mother substance	Daughter substance	Half-life (millions of years)
Uranium 238	Lead 206	4,510
Uranium 235	Lead 207	713
Thorium 232	Lead 208	13,900
Potassium 40	{ Argon 40 { Calcium 40	11,850 1,470
Rubidium 87	Strontium 87	50,000

uranium atoms that have turned into lead we can tell how much time has elapsed. Bear in mind that what is constant is the half-life, that is, the time required to transform half the uranium that is left each time. The half-life of uranium 235 is 713 million years. Given 100 grams of uranium 235 it takes 713 million years for half this amount, or 50 grams, to turn into lead 207; in the next 713 million years the amount that turns into lead is not the remaining 50 grams, but only 25 grams; and in the next 713 million years it will be half of the remaining 25 grams; and so forth. Due to this law of radioactive decay, even a small amount of uranium is sufficient to measure very long periods of time. Suppose we find a rock that while forming incorporated a certain amount of uranium. From the amount of lead found in it at present, which had to be uranium originally, we can calculate how much time has elapsed, that is, how many years ago the rock formed and incorporated uranium.

This method also works for other radioactive, or unstable, elements. Table 1 lists the most important radioactive isotopes for dating rocks and strata, along with their respective half-lives and the elements they turn into.

Radioactive dating has enabled us to estimate the ages of all the events recorded in the great book of rock strata, which in the past could tell us only what came before and what came after. Later on it will help us to trace the evolution of the earth from its origin to the present. Right now we shall use it to go further and further back in time and to determine the age of the earth.

As we date ever more ancient objects we discover that beyond a certain time we find nothing but rocks. The oldest rocks have been found in Greenland and are 3.75 billion years old. It is possible that yet older rocks may exist, undiscovered, but we certainly shall never find rocks that formed at the same time as the earth and remained unchanged down to the present. Hence we cannot discover when the earth was born by dating its rocks.

The age of the earth has been established in a different way, namely, by determining the age of celestial bodies like the moon and the meteorites that we have reason to believe formed at the same time as our planet. In this manner we have found that the earth was born 4.6 billion years ago. Naturally at that time the earth's crust did not yet exist, and that is why we shall never find rocks of that age on the earth. Perhaps there were only dust and solid fragments in the process of aggregating, which would later fuse; perhaps there was only molten material. Certainly the planet earth as we know it today did not exist.

In any case, the earth, the moon, and the solar system in general are not eternal; they were formed in a very remote but well-defined epoch. And the changeableness of the world that we had begun to recognize from the instability of the cycles and their basic unreality reaches the point of rendering unreal even things that today exist in such a solid and obvious way as to give us the feeling that they have always existed. Of course, there was something else in place of the solar system—the matter from which it formed, for example, and this matter perhaps existed from the beginning of time, or perhaps had itself just formed. One thing is certain: the solar system, as such, originated a little over 4.5 billion years ago. Before that time the solar system, and hence the earth, did not exist.

THE ORIGIN OF THE UNIVERSE AND OF THE EARTH
In our first quick look at the universe we discovered that the galaxies are running away from one another. This observation leads to the conclusion that the entire universe is expanding. It follows that as we go back into the past the universe gets smaller and smaller and the galaxies move closer and closer together. By observing the galaxies today, we can calculate the time in which all matter was jammed within an exceedingly small space, perhaps a point. That is when the universe began, and we can assume that it happened 15 billion years ago.[3] It began

with a colossal explosion, known as the Big Bang, which gave rise to space and time.

The discovery of the Big Bang proves that not only the earth but the entire universe had a beginning. Furthermore it establishes that the universe is at least 15 billion years old. But in the past few years we have found that the earth, along with the solar system, formed less than 5 billion years ago. Thus there was a very long period, about twice the time elapsed from the formation of the earth to the present, during which the universe existed with its millions of galaxies and billions upon billions of stars, but the earth and the other planets did not. What was in their place? Was the rest of the universe the same as today? And did the delay in the earth's birth have any effect on its history and particularly on the human beings that inhabit it?

Today, at long last, we can begin to answer these and many other questions.

2 ORIGIN, LIFE, AND END OF THE STARS

FUNDAMENTAL CHARACTERISTICS

To find out how the sun and the earth were formed, we must first learn how stars in general are formed. And if we can also discover how they evolve and end, we shall be able to reconstruct the entire life of the sun and predict its future. Furthermore, by studying the celestial bodies in our neighborhood we shall get an idea of how the universe works, at least the small part of it that we can most readily see and investigate. Unfortunately, the span of human life is very brief compared with that of the stars. Their evolution is so slow, in fact, that nothing in the world inspires so great a feeling of eternity as the celestial vault. Whether the stars change is something that we cannot hope to discover in one lifetime or through centuries of observations handed down by generations of astronomers. There is another way, however.

Imagine an extraterrestrial being who, knowing nothing about human life, lands in one of our cities and sees its inhabitants—children, youths, adults, and old people. He wishes to discover whether these creatures always remain the same, like the Greek gods on Mount Olympus, or whether they are apt to change in the course of time and especially whether they have a beginning and an end.

It would not be difficult for him to learn the truth if he could spend several years among the people of the city, even if most of them lived to a hundred. But our extraterrestrial is pressed for time and must complete his investigation in less than a day utilizing only the research tools of modern biology and a keen brain. He then seeks to study the human body, to explain its functions. He soon discovers metabolism and the fact that processes of assimilation and synthesis prevail in young people, while processes of disintegration and decay characterize old people. If he is lucky enough to visit a maternity ward, he can assist at the birth of an infant and observe other creatures that by their generic similarity to the infant he assumes must also be newborns. In a nursing home he might attend the death of an old man and could see and study the other organisms, from which he would get useful information about the last stages of human life. In this manner he will be able to leave for his planet with a pretty accurate view of our life cycle despite the fact that he has not been able to follow the entire life of a single individual.

This is more or less the way astronomers proceed. But the stars live far longer than our centenarians. The sun is more than 4.5 billion years old and will endure for at least as many years. Even assuming that the stars lived only a billion years, if you consider that stellar observations have been carried out mainly in this century, it is obvious that they comprise but a tiny fraction of a star's life. It is as if our extraterrestrial had a total of 5.2 minutes to discover that the people of the city are born, live, and die.

Nevertheless we are going to try, and it will be worth our effort because the view we shall obtain is far more rewarding than a starry sky, frozen in its splendor. It is a view that we have to conquer little by little, by observing and reasoning, much as a mountaineer discovers an ever wider and more distant panorama as he laboriously climbs toward a summit. But the panorama that we shall see with our mind's eye will be a changing one, showing us what happened in the past when we did not yet exist and what will happen in the future when we shall be long gone—a perspective that beside letting us rove through the cosmos will extend us in time as if we had enormously increased the span of our lifetime.

BASIC DESCRIPTORS OF THE STARS
Let us start with a simple reflection: The stars *must* change with the passage of time. The reason is simple. A star shines because it generates and emits energy in the form of light, heat, and other kinds of radiation. And just as the wood burning in a stove eventually turns to ashes, something of the same kind must also happen in the stars. Hence, what we must find out is not if a star changes but how it changes.

Every star has some fundamental chemical and physical characteristics such as mass, luminosity, chemical composition, surface temperature, and (higher) internal temperature. In the course of a star's life all these characteristics change, some more some less. There are variations in mass and chemical composition, but they are difficult to observe— the mass because it can only be determined for some double stars; the chemical composition because its changes occur mainly in the interior of the star, which is hidden from us. What we can more readily obtain from observations are a star's luminosity and surface temperature, which are also the physical quantities that must change the most.

Actually, observations do not immediately yield temperatures and luminosities, but rather spectral types (or colors) and absolute magnitudes, from which the first two can be derived. The theorists, on the other hand, starting from actual or theorized physical quantities arrive directly at luminosities and temperatures. In both cases the results may be illustrated, either by a diagram in which spectral types or color indices (both of which can be expressed by numbers) are plotted along the abscissa and absolute magnitudes are plotted along the ordinate or by a diagram that directly plots temperatures (along the abscissa) versus luminosities (along the ordinate). In either diagram a star of a given temperature and luminosity is obviously represented by a point. A star of different temperature and luminosity will be represented by another point located in a different position. If with the passage of time a star undergoes changes in temperature or luminosity or both, the point representing the star will move to a different position (figure 10). A diagram constructed for a group of stars of different temperatures and luminosities is known as a Hertzsprung-Russell diagram, or H-R diagram for short.

THE HERTZSPRUNG-RUSSELL DIAGRAM

At the beginning of the century E. Hertzsprung and H. N. Russell independently constructed this type of diagram for all the stars whose spectral types and absolute magnitudes were known. What they found (see figure 11) was that the stars did not all fall at the same point, which meant that they were not all equally hot and equally bright. On the other hand, they were not scattered all over the diagram, as they would have been if they could have any temperature and luminosity whatever. Instead, they tended to be concentrated in well-defined zones. Most of the stars fell in a narrow belt running diagonally across the diagram, called the *main sequence*. Another large group of stars fell above the belt and to the right. The stars of this second group have the same spectral types and hence the same (low) surface temperatures as the fainter stars of the main sequence, but are much brighter. By a well-known physical law (Stefan's law) stars of the same temperature must emit the same amount of radiation per unit of surface area; in other words, a square meter of the fainter star emits the same amount of light as a square meter of the brighter star. Hence the difference in brightness between the stars of the two groups must be due solely to a difference

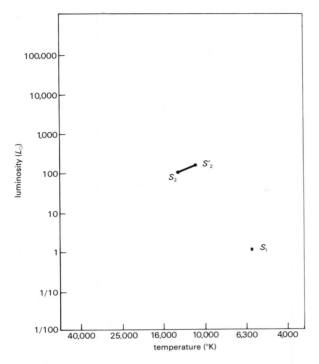

Figure 10   Two stars in the temperature-luminosity diagram. The sun's luminosity, $L_\odot$, is used as the unit of luminosity. The unit of temperature is 1°K (1 degree Kelvin). The point S_1 represents a star with the temperature and luminosity of the sun, while S_2 indicates a star much hotter and more luminous. If in time this star should become more luminous but cooler, its representative point would shift—for example, from S_2 to S_2'.

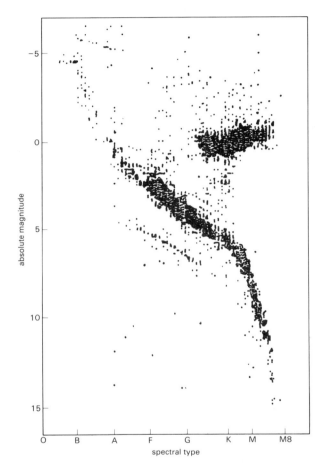

Figure 11   The Hertzsprung-Russell diagram, with each point representing a star. The great majority of stars fall in a belt running diagonally across the diagram from top left (hot, blue, luminous stars) to bottom right (cool, red, faint stars). Note the group of red giants above the main sequence. [From *Astronomičeskij Žurnal*]

in surface area. It follows that the fainter stars of the main sequence must be much smaller than the bright stars of the second group. The former were called *red dwarfs* and the latter *red giants*.

The Hertzsprung-Russell diagram also showed stars that were even brighter than the red giants, named *supergiants*, and stars that were as faint as the red dwarfs but generally white in color, called *white dwarfs*. Supergiants and white dwarfs seemed to comprise only a small minority of the stars, but one could not really be sure that every group in the diagram was truly representative of the actual percentage of such stars in space. The diagram had been constructed with all the stars for which spectra could be obtained and distances could be measured; these were generally the brightest stars. As a result the diagram was biased in favor of the stars that were intrinsically more luminous. It was entirely possible that the red giants appeared as a conspicuous group simply because they could be easily seen even at great distances, while the white dwarfs seemed few because they were so faint that only the nearest ones could be seen. In fact, if we remove this selection effect as much as possible —for example, by considering only the stars within a certain distance from us—it turns out that 85% of the stars belong to the main sequence, 3–6% are white dwarfs, and the remaining 10% are giants, supergiants, and other types of stars.

The H-R diagram is the observational key that enables us to solve the problem of stellar evolution. We have established that the temperature and luminosity of a star must change in the course of time—recall the burning log! We also have seen that any such change is reflected in a displacement of the star on the temperature-luminosity diagram (in practice, on the H-R diagram). Thus, while it is true that the points on the diagram correspond to the positions occupied by different stars at a given time, these points also correspond to the different positions that the same star occupies during its life as its temperature and luminosity change. Consequently, to learn how the stars evolve we must find out how they move across the diagram, that is, which points they occupy at different ages and in which order. Perhaps the stars are born as giants and then become dwarfs, or perhaps it is the other way around. Also, there may be stars whose temperatures and luminosities are different for other than evolutionary reasons and remain unchanged, if not for all eternity, at least for very long periods of time. All these problems have been satisfactorily resolved only in the past few years. In the process, besides learning how the stars evolve, we have discovered how the elements are built, how we have come to the present world, and how the universe has evolved up till now.

FORMATION OF THE STARS

In interstellar space there are large amounts of gas and dust that can be seen as bright nebulae if they are illuminated or excited by a nearby star or as dark nebulae if no star illuminates them but they are projected on a rich stellar field or a bright cloud. This interstellar matter consists of almost 80% hydrogen and almost 20% helium, with all the other elements comprising the small remaining percentage. It is from this gas and dust that the stars are formed.[4] They are generally born in groups when interstellar matter accumulates, breaks up into fragments, and then condenses within each fragment. This can occur in different ways, which we shall not go into now. The important thing is that these contracting fragments exist and have been observed, and this being the case, we prefer to see immediately what happens. With these observations that show us, not the birth of the stars, but the gathering of the material that is about to form them, we embark on an adventure during which we shall witness the origin, the life, and the end not only of the stars but also of the planets and of one of the many worlds, the earth, in which we ourselves enter as transient but conscious protagonists.

THE DARK GLOBULES
In 1947 B. J. Bok and E. F. Reilly called attention to a particular type of dark nebulae, very dense and round in shape, known ever since as "Bok's globules," or simply "globules." They are of two types: smaller globules that can be seen projected against a bright nebula (figure 12) and larger globules that appear as black round spots on the background of the starry sky (figure 13). At first both types of globules were thought to be the clouds of gas and dust that by contracting and heating up burst into brilliance as stars, but according to G. H. Herbig this is not the case with the smaller globules. On the other hand, recent observations with optical and radio instruments have confirmed that the larger globules are indeed cold dark clouds in gravitational collapse, clouds, that is, in which all the material is falling toward the center under the pull of gravity.

Figure 12   Dark clouds in the Rosette nebula. Observe that some of them are very small and nearly round. These dark globules are believed to be clouds in gravitational contraction, the initial phase of star formation. [Courtesy Mount Palomar Observatory]

Figure 13   A dark globule, Barnard 335, visible as a dark starless region (center top) on the background of a stellar field. Photograph made with the 2.25-meter telescope of Steward Observatory. [Courtesy B. J. Bok]

Studies of eight such globules show that they have very low temperatures, about 10°K, radii ranging from 1 to 3.8 light-years, and masses ranging from 19 to 750 solar masses. It is worth mentioning that a globule of about 20 solar masses can produce only a single, though massive, star, whereas the exceptionally large globule of 740 solar masses will certainly lead to the formation of a group of stars.

Let us consider the case of a single star. In the collapsing cloud the energy due to the motion of the in-falling material is converted into heat. The temperature of the globule increases, especially in the central region, causing the dust to evaporate and reducing all the material to the atomic state. As the atoms approach the center of the cloud the pressure also increases until it becomes high enough to balance the pull of gravity and halt the fall of the overlying material. At this point the collapse stops. A protostar is born.

This phase is relatively rapid, at least compared with the entire life-time of a star; it lasts a few hundred thousand years for the stars of larger mass and a few million years for the others.

THE PROTOSTARS

The collapse has stopped and the object has begun to glow, even though we do not see it since the temperature is high only near the center, which is still surrounded by a cocoon of dust; but it is not yet a bona fide star because there is no production of nuclear energy in its interior. Although the collapse has stopped, energy continues to be generated by gravitational contraction, which, however, now occurs much more slowly. According to a relation known as the virial theo-rem, half this energy is emitted outward and half heats the interior. Hence with the passage of time the temperature gradually rises and the cocoon of dust is pushed away and dispersed in space. As the obscuring envelope dissolves the protostar becomes visible.

As the protostar keeps on contracting, its internal temperature rises until it becomes high enough to ignite thermonuclear reactions.[5] To put it simply, the two main sets of reactions that occur in this phase are the "proton-proton" cycle and the "carbon-nitrogen" cycle. The first occurs at temperatures between 10 million and 16 million degrees; the second, at temperatures between 16 million and 30 million degrees. In both cases hydrogen is converted into helium with a 7‰ loss of mass that turns into energy according to Einstein's celebrated formula

$$E = mc^2,$$

where E is the energy resulting from the annihilation of a mass m and c is the speed of light. At this point the production of energy in the inte-rior of the protostar has increased to the point of arresting the gravita-tional contraction and the object assumes a stable configuration that will endure until all the hydrogen in its interior has turned into helium, that is, for millions or billions of years, depending on the circumstances. Now we finally can say that a star is born.

By expressing the various stages of star formation in quantitative form, the theorists have been able to determine how the surface tem-peratures and luminosities of the forming stars change in time and thus to demonstrate how the protostars move on the temperature-luminosity diagram. It turns out that the path of a protostar depends on its mass. Let us consider a star whose mass is one and a half times that of the sun. At first the luminosity diminishes because the star contracts a great deal but the temperature and emission of energy per unit of surface area increase only a little. Then the temperature rises faster, the contraction slows down, and the star's luminosity again begins to increase. Figure 14 shows how, according to calculations by I. Iben Jr., temperature and luminosity vary for stars of masses ranging from $0.5M_\odot$ to $15M_\odot$ ($M_\odot$ = solar mass); table 2 gives the times required to travel the various tracks. Note that the period of gestation runs from 155 million years for stars of 0.5 solar mass to 62,000 years for stars of 15 solar masses.

What marks the true beginning of a star and characterizes its "normal" life is its relative stability. The protostar stage, in fact, is not as calm and straightforward as it might appear from my description and from the diagrams derived from theoretical calculations. Bear in mind that this is a period in which everything is continuously changing and everything is in the process of forming, from the structure of the star to a possible planetary system, which at this point would still be in the shape of a dust envelope or disk surrounding the star.

This is amply borne out by observations. Astronomers have observed stars still immersed in their cocoon of dust and perhaps stars that are about to emerge from it. At this stage nothing can be seen because the dust envelope absorbs the light and the ultraviolet radiation emitted by the protostar inside. However, the energy absorbed by the envelope is transformed and reemitted mainly as infrared radiation, which can be detected.

In the late 1960s infrared sources began to be observed in the regions where stars are likely to form, such as the Orion nebula, and from the first they were thought to be cocoons containing protostars or even hot stars. One of them was the Becklin-Neugebauer infrared source, named after the astronomers who discovered it in 1967, and known today as the "B-N source." In 1978, through infrared spectroscopic observations performed with the 4-meter telescope at Kitt Peak, D. B. Hall and his associates found that in the center of the B-N source there is a star of type B0 or B1 (that is, very hot) surrounded by a compact region of ionized hydrogen and a larger region of hot dust. They also discovered an extended circumstellar cloud of carbon oxide that is expanding, blown away by the stellar wind that in the opinion of the

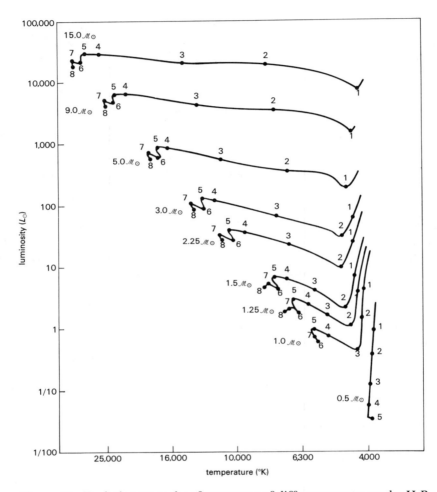

Figure 14   Evolutionary paths of protostars of different masses on the H-R diagram. The mass is indicated on the left of each path. The numbered points refer to various stages in the calculations; the corresponding times are listed in table 2.

Table 2   Time elapsed from the initial model for protostars with masses ranging from 15 to 0.5$M_\odot$

Position of the stars on the paths drawn in figure 14	15$M_\odot$ (tens of years)	9$M_\odot$ (hundreds of years)	5$M_\odot$ (hundreds of years)	3$M_\odot$ (thousands of years)	2.25$M_\odot$ (thousands of years)	1.5$M_\odot$ (tens of thousands of years)	1.25$M_\odot$ (tens of thousands of years)	1.0$M_\odot$ (tens of thousands of years)	0.5$M_\odot$ (hundreds of thousands of years)
1	67	14	294	34	79	23	45	12	3
2	377	15	1,069	208	594	236	396	106	18
3	935	364	2,001	763	1,883	580	880	891	87
4	2,203	699	2,860	1,135	2,505	758	1,115	1,821	309
5	2,657	792	3,137	1,250	2,818	862	1,404	2,529	1,550
6	3,984	1,019	3,880	1,465	3,319	1,043	1,755	3,418	—
7	4,585	1,915	4,559	1,741	3,993	1,339	2,796	5,016	—
8	6,170	1,505	5,759	2,514	5,855	1,821	2,954	—	—

a. Adapted from I. Iben Jr., *Astrophys. J.* 141 (1965):993.

discoverers has been blowing for less than 1,800 years. Consequently, rather than a protostar still in the phase of gravitational collapse, the B-N source appears to be a young star that for the last 1,800 years has been ridding itself of the cocoon, the residue of the material from which it formed, which still prevents us from seeing it directly.

Another very interesting case is the extraordinary object known as η Carinae, possibly a whole group of forming stars that sometimes can be seen shining through the cocoon, which is beginning to break up.[6]

Turning to the protostars that are visible optically, their strange and diverse behaviors reflect well the chaotic conditions prevailing at this stage. While generally confirming our views, however, they also show that when it comes to details many problems remain to be solved. And perhaps it is more than just details. Many of these objects (the variable stars of the T Tauri type) have anomalous spectra that reveal the presence of gaseous envelopes and exceptionally high amounts of lithium. Many others fluctuate irregularly in brightness; some appear to be accreting material from the outside, and a large number of them have sudden flare-ups that do not last much more than a quarter of an hour.

The latest phenomenon to be observed in protostars and certainly one of the most puzzling is known as "fuor." According to Herbig, it could well play a major role in the formation of planetary systems. Let me describe it briefly.

Between 1936 and 1937 there appeared a star in Orion that in the following years declined slightly in brightness and then remained nearly constant (figure 15). The new object, initially known as Wachmann's star after the astronomer from Hamburg who had discovered it, was later listed in the General Catalog of Variable Stars as FU Orionis. It was certainly a variable star because its brightness had increased enormously. But what sort of a variable? The fact that after a couple of years at maximum it had begun to fade suggested a slow nova, but this idea had to be rejected when its brightness subsequently was observed to remain constant. Meanwhile our knowledge of stellar evolution improved and the object was found to be located in a region where stars are forming and to have a spectrum similar to that of a T Tauri star. As a result it was thought that we had seen the formation of a T Tauri star, the very moment in which the gravitational collapse leads to the kindling of a protostar.

A similar case occurred about 1970, but this time the event was stud-

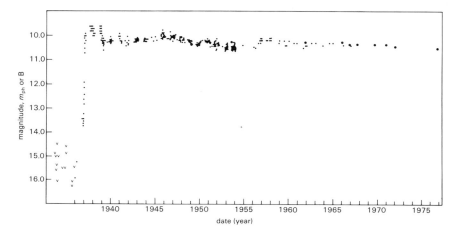

Figure 15 Photographic light curve of the star FU Orionis from 1936 to 1977. The differently shaped points partly indicate that the data were obtained by different astronomers using different instruments. The V-shaped signs indicate that at the time of the photograph the star was fainter than the magnitude corresponding to the sign. [Adapted from G. H. Herbig, *Astrophys. J.* 217:694, © The University of Chicago Press]

ied much more effectively. A star in the constellation Cygnus, later denoted V 1057 Cygni, suddenly flared up, brightening by 6 magnitudes (250 times) in 400 days (figure 16). At maximum its spectrum was similar to that of FU Orionis. But as luck would have it, a spectrum of the star had been obtained 12 years before, and it showed that at that time it was already a T Tauri star. It follows that the flare-up does not mark the kindling of a protostar but, for reasons as yet unclear, occurs in protostars that are already kindled. This was confirmed in 1974 with the discovery of a third case—V 1515 Cygni, a star that has been steadily brightening since 1949 and whose spectrum is similar to that of the other two.

In all three cases, moreover, with the rise in brightness a small, arc-shaped nebulosity appeared near the star, though not centered on it. This nebulosity could not have formed from the material ejected by the star during the flare-up. Spectroscopic observations of V 1057 Cygni show that the star did indeed eject a gaseous envelope and perhaps more than one, first at a velocity of 500 km/sec and then of 300 km/sec. But the observed nebulosity is stationary with respect to the star. In

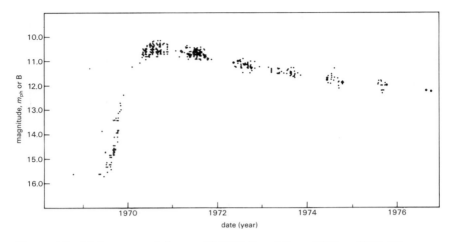

Figure 16   Light curve of the star V 1057 Cygni from 1968 to 1977. The symbols are the same as in the previous figure. [Adapted from G. H. Herbig, *Astrophys. J.* 217:696, © The University of Chicago Press, by permission of Lick Observatory]

addition, photographs taken in the following years show that as the star declined in brightness the nebulosity faded away (figure 17). This indicates that the nebulosity was already there before the flare-up and became visible only when the star brightened enough to illuminate it. Almost certainly it formed during a previous explosion when the ejected gas swept away the dust surrounding the protostar and formed the ring that is now centered on the position occupied by the star at that time.

For this strange phenomenon the Soviet astronomer V. Ambartsumian proposed in 1971 the name "fuor" from the first four letters of FU Orionis. Its causes are not known, but it appears to be a sporadic flare-up accompanied by the ejection of one or more gas envelopes at high speed. In all likelihood it happens more than once in a certain period of a protostar's life. G. H. Herbig, the astronomer who has studied this sort of phenomena more than anyone else, has calculated that it should occur once every 10,000 years. He also has found that three other variables (EX Lupi, VY Tauri, and UZ Tauri) are stars of the T Tauri type and display similar light phenomena, with less intense but much more frequent flare-ups. This suggests that a similar activity

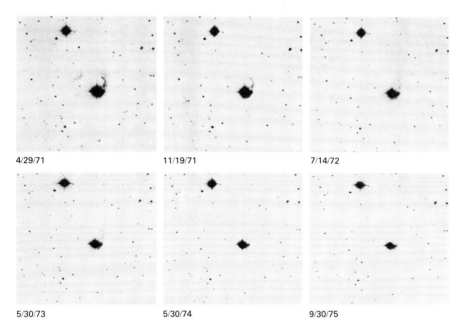

Figure 17   The star V 1057 Cygni and its surrounding nebulosity. The decline in brightness of the variable star appears clear from a comparison with that of the star at top left. Observe that as the variable gets dimmer the nebulosity that reflects its light becomes less and less visible. These photographs were obtained from 1971 to 1975 with the Crossley telescope of the Lick Observatory. [Courtesy G. H. Herbig]

occurs in T Tauri stars of different ages, becoming less intense but more frequent as time goes by, as if it was abating. With the description of this phenomenon we have concluded our observations of stars in the process of forming. We have seen them ignite. Let us see how they live.

THE MAIN SEQUENCE

As I mentioned earlier, a protostar becomes a true star when the thermonuclear reactions begin and the object reaches a condition of equilibrium, that is, when it neither collapses nor expands.

Let us now consider a star and assume that it is in equilibrium, that it has a certain chemical composition, and that it has a certain mass; in this case, as H. Vogt and N. H. Russell independently demonstrated in

1926, both its temperature and luminosity must have a specific value. In other words, the mass and chemical composition of a normal star determine precisely its position on the H-R diagram. Let us consider now the stars in a cluster, and more precisely the stars of a young cluster like the one embedded in the Orion nebula, where many stars are still forming. Since they were all formed from the same material, these stars must have the same chemical composition. Hence if they were all born with the same mass, they would all become stars of the same temperature and luminosity; in other words, once they had reached equilibrium after the protostar phase, they would all be represented by a single point on the H-R diagram. But this is not what we find when we plot the diagram on the basis of observations; most of the stars fall in a belt that, if you recall, is known as the main sequence (figure 18). Since all of these stars have the same chemical composition, it follows that their temperatures and luminosities (and hence their positions on the diagram) depend solely on the mass. Thus the various points on the main sequence correspond to stars of different mass.

Figure 18 also shows a number of points scattered above the main sequence and becoming more numerous as one moves to the right. These points represent stars that have not yet reached equilibrium; in other words, they correspond to protostars rather than stars. This has also been demonstrated theoretically. As figure 14 and table 2 show, the more massive the star, the faster it will reach the main sequence. The time required to get there is 50,000 years for a star of 15 solar masses, 500,000 years for a star of 5 solar masses, 34 million years for a sunlike star, and as much as 155 million years for a star of 0.5 solar mass. Turning back to the diagram in figure 18, the scattered points now are easily explained. Since protostars of larger mass spend relatively less time outside the main sequence, it is obvious that as we move up the sequence we find fewer and fewer of them above the belt. On the other hand, the many points scattered on the right correspond to less massive protostars, which, requiring more time to arrive at the main sequence, have not yet reached it and are still in the stage of gravitational contraction.

One additional fact must be noted. According to the Vogt-Russell theorem, the stars of the same chemical composition should be distributed in a line rather than a belt. In effect the belt of the main sequence includes stars that are about to reach it but have not yet kindled ther-

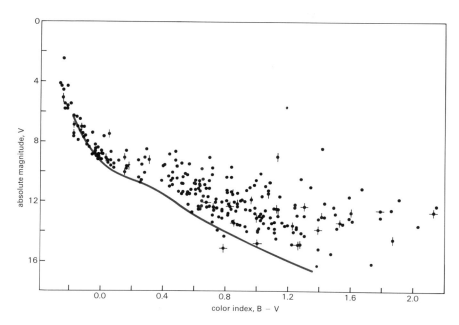

Figure 18   The main sequence of the H-R diagram plotted by M. F. Walker
for stars of the association in Orion. Most of the points fall above the zero-age
line because the corresponding stars are still in the phase of gravitational contraction that precedes the start of the first thermonuclear reactions. The effect
becomes more conspicuous the farther we move to the right, where red stars
less massive than the sun are located, confirming the theoretical predictions
(see figure 14). In this figure, as in analogous diagrams, the color index (horizontal axis) indicates the temperature of the stars, which increases as the value
of the color index decreases. The color index used here, B − V, is defined as
the difference between the star's blue (B) and visual (V) magnitudes. The apparent (*m*) or absolute (*M*) magnitude is plotted on the vertical axis.

monuclear reactions, as well as adult stars that, like all adults, begin to show some changes with the passage of time. In this case the surface temperature and luminosity are beginning to change. The conditions of equilibrium postulated by Vogt and Russell certainly obtain when the thermonuclear reactions begin. Hence stars in this stage fall on a line, called the zero-age line, which borders the lower edge of the main sequence.

EVOLUTION OF THE STARS

The star has formed. It is on the main sequence, in equilibrium, and radiates heat and light by burning its nuclear fuel.[7] Clearly, this state of affairs cannot last forever: The production of energy and the state of equilibrium eventually must end, at least when the nuclear fuel is exhausted. At this point, one way or another, the star must come to an end.

The most significant clues to stellar evolution are furnished by the H-R diagrams for clusters. These diagrams are easier to interpret because the components of each cluster constitute a homogeneous group; having originated from the same material, they can be assumed to have the same chemical composition. In any case the homogeneity of the composition can be ascertained spectroscopically cluster by cluster. Disregarding the clusters that do not have this prerequisite, one may compare the H-R diagrams of clusters composed of stars with the same chemical composition. Given the same mass distribution for all of these clusters, their H-R diagrams should all be the same. But this is not what we find. Why? The answer to this question is the key to stellar evolution. In addition, as we shall see later, the comparison of different H-R diagrams, corresponding to clusters that show more or less marked differences, also will yield the key to understanding the evolution of the universe as a whole.

CLUSTER STARS

Figure 19 shows the H-R diagrams for the clusters h and χ Persei, M 11, Praesepe, and M 67. In all four diagrams the lower part stops at the faintest stars that could be seen with the telescopes used in the observations of the clusters. In addition, each diagram contains a number of field stars that seem to belong to the cluster because they are on the

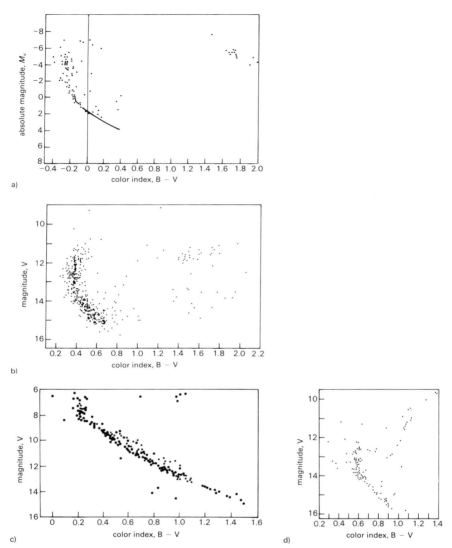

Figure 19 H-R diagrams for four galactic clusters: (a) h and χ Persei (here the color index is plotted against the absolute magnitude, and the heavy line indicates the zero-age line); (b) M 11; (c) Praesepe; (d) M 67. [(a) from H. L. Johnson and W. A. Hintner, *Astrophys. J.* 123:267, © The University of Chicago Press; (b) from H. L. Johnson, A. R. Sandage, and H. D. Wahlquist, *Astrophys. J.* 124:84, © The University of Chicago Press; (c) from H. L. Johnson, *Astrophys. J.* 116:640, © The University of Chicago Press; (d) from R. Racine, *Astrophys. J.* 168:399, © The University of Chicago Press]

line of sight; in effect they are not a part of it and fall out of place with respect to the cluster's diagram. Despite these limitations there are obvious differences among the four diagrams.

Praesepe clearly shows the main sequence and 4 red giants. In h and χ Persei the main sequence extends to very luminous stars and then veers slightly to the right; this double cluster has a fair number of red giants. M 11 has a more conspicuous group of red giants and its main sequence also veers to the right, but at redder stars. Finally, M 67 is different from the other three in that the main sequence is short and continuous with the branch of the giants.

These differences become more significant if we compare the diagrams for clusters with the absolute magnitudes instead of the apparent magnitudes. This can be done even if the distance is not known for all of the clusters. According to the Vogt-Russell theorem, and as confirmed by observations of the nearest clusters, the zero-age line, which marks the lower edge of the main sequence, must be the same for all the clusters. Once we have plotted the H-R diagram with the colors and absolute magnitudes of a cluster of known distance (as here in the case of h and χ Persei), we only have to slide all the other diagrams so that the main sequence of each cluster, or at least the observed piece of it, coincides with that of the first. In this manner in 1957 A. Sandage constructed a combined H-R diagram for ten clusters (figure 20).

This diagram revealed three very important facts, whose interpretation furnished the observational key to our understanding of stellar evolution. First of all observe that the top of the main sequence of each cluster extends to a different value of the absolute magnitude. In other words, the brightest stars are not equally bright in all the clusters; for example, the brightest stars in NGC 2362 are 10,000 times brighter than the brightest stars in M 67. To put it another way, the main sequence of many clusters lacks the brightest stars, which are also the most massive. In the second place, notice that the main sequence of each cluster at the point where it stops veers to the right toward a group of stars, also belonging to the cluster, that are red giants. Finally, in nearly all the clusters there is a gap, more or less wide, between the curving part of the main sequence and the branch of the red giants, while in other clusters the main sequence joins the giant branch in a

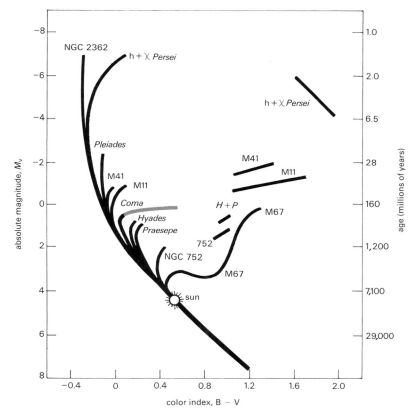

Figure 20   The H-R diagrams for ten galactic clusters have been combined by superimposing their zero-age lines, which we assume to be all the same. The ages of the clusters, shown on the right-hand side, are obtained from the turn-off points. [Adapted from A. Sandage, *Astrophys. J.* 125:436, © The University of Chicago Press]

continuous line. As I said, these results furnish the observational key to
the problem of stellar evolution. Let us now interpret these findings.

We have seen earlier that main-sequence stars are stable because they
produce energy by converting hydrogen into helium. As M. Schönberg
and S. Chandrasekhar found theoretically in 1942, the more luminous
and massive the star, the faster it will burn its hydrogen; and when a
certain percentage of hydrogen has turned into helium the star leaves
the main sequence. According to their calculations, this occurs when
the burned-up hydrogen is 12% of the total mass. At that point the
star becomes more luminous and less blue and after passing rapidly
through a period of nonequilibrium becomes a red giant. This theory
explains the curving upper part of the main sequence, which consists of
the stars that are departing from it, as well as the group of red giants
and the gap in the middle. Further confirming Schönberg's and Chandrasekhar's predictions, a small number of stars do fall in the gap, and
they turn out to be unstable. These are the Cepheid variables, huge
globes of gas that reveal their instability through periodic pulsation.[8]

On this basis we can begin to explain the H-R diagrams for clusters.
At first all the stars are on the main sequence, except for the stars of
small mass, lower right, that have not yet reached it (figure 18). In time
the more massive stars depart from it and become red giants. The
brightest stars are the first to leave, but little by little they all do, with
the less massive stars spending the longest time on the sequence. Hence
from the point at which the main sequence turns right toward the region of the red giants it is possible to calculate the age of the cluster,
shown in figure 20 on the right-hand side.[9] After leaving the main
sequence, the more massive stars go through a period of instability
that, compared with the entire lifetime of a star, lasts no longer than
the time to jump a ditch. During this stage they pulsate as Cepheid
variables. Upon becoming red giants they enter a phase of lower instability. Many of them still pulsate, but more slowly than the Cepheids
and in a different way; such are the variables of the Mira Ceti type and
the semiregular variables.

The theoretical results of Schönberg and Chandrasekhar, however,
did not explain all the H-R diagrams for clusters. In M 67 and NGC
188 the main sequence branches off in a continuous line and there is
no region of instability. Furthermore, as figure 20 clearly shows, this
region of instability is smaller where the cluster consists of stars of small

mass; the smaller the masses, the more the gap shrinks. This can mean only one thing: Stars evolve in different ways depending on their masses, not only in the protostar phase, as we saw earlier, but also after reaching the main sequence.

To clarify this point we must understand the physical changes that cause the stars to follow different paths on the H-R diagram. There is only one way to do this: explore their interiors and observe how they change in time.

STELLAR MODELS

Unfortunately, the astronomer cannot observe the interior of a star for the obvious reason that it is hidden by the overlying layers. The problem can only be solved theoretically by means of stellar models whose validity ultimately depends on how well they fit the observations. All such models must fulfill a basic requirement—namely, they must satisfy the fundamental laws of physics and the conditions of equilibrium.

To construct a stellar model one starts by assuming that the star has spherical symmetry (that is, that its state varies only with the distance from the center) and by presupposing two fundamental principles and two conditions.

The two principles are *conservation of mass* and *conservation of energy*. The first is well known and to a first approximation it can be assumed that the star satisfies it. The second does not appear to be satisfied since the surface of the star is continuously dissipating a huge amount of energy. If this principle is to be satisfied—and it must be, or the star would soon be extinguished—an amount of energy equal to that radiated outward must be continuously produced in its interior. Mathematically speaking, this principle is translated into an equation that gives the luminosity at various depths as a function of the production of energy at those depths.

Let us consider the two conditions in turn. First of all, in order for a star to be stable, in each of its layers the weight of the overlying gas must be balanced by the outward pressure from below. This condition is known as *hydrostatic equilibrium*. If those forces do not balance each other, the star becomes unstable; it contracts when the weight of the overlying layers is greater than the pressure from below, and it expands when there is more pressure from below.

The second condition is the *transport of energy*. It requires that the heat, i.e., the energy, produced in the interior be conveyed outward. Let us see how this happens in practice.

Heat propagates by three familiar mechanisms: conduction, convection, and radiation. Conduction is the mechanism of energy transport at work in solids. As you probably know by experience, if you leave one end of a spoon in a cooking pot, the other end will get hot enough to burn your hand. Convection is the process whereby heat is carried by matter in mass motion. The kitchen will again provide us with a good example. In a pot of boiling water the liquid closer to the bottom, and hence to the stove, heats up and rises to the surface; heat is thus carried by the upward flow of the liquid. Coming in contact with the cooler air outside the pot, this liquid cools down and sinks back to the bottom. Finally heat can be transported by radiation, that is, electromagnetic waves. This mechanism works not only in vacuum but also through matter provided it is "transparent," much as a transparent window pane allows the warm and luminous rays of the sun to come through.

In the interior of the stars energy is normally transported by convection and radiation, but in some cases, as we shall see later, it is transported almost exclusively by conduction.

Let us see now how a star model is calculated and how the evolution of a star may be traced by means of a series of models. To begin with, the theoretical astrophysicist divides the star into a number of concentric shells. For each of these shells he writes four equations that express in mathematical terms the two principles and the two conditions stated above. By solving the four equations he obtains the temperature and pressure in each shell, the mass it contains, and the energy it emits. Figure 21 shows the results obtained for the interior of the sun.

With these data one can understand the internal structure of the star and identify, in the regions hot enough for thermonuclear reactions to occur, the type of reaction that produces the energy. Naturally the star must be divided into a fairly large number of shells, say 100. It follows that to calculate a model of the star at a specific time one has to solve at least 400 equations. In the past this required a large number of long and tedious calculations. In the 1950s, for example, it would take a researcher a year of work to calculate a model of a star, and ultimately all he had was a description of the star at a given moment. In order to follow the evolution of the star he had to calculate at least 10 models

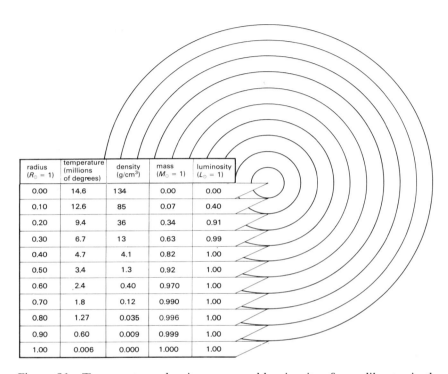

radius ($R_\odot = 1$)	temperature (millions of degrees)	density (g/cm^3)	mass ($M_\odot = 1$)	luminosity ($L_\odot = 1$)
0.00	14.6	134	0.00	0.00
0.10	12.6	85	0.07	0.40
0.20	9.4	36	0.34	0.91
0.30	6.7	13	0.63	0.99
0.40	4.7	4.1	0.82	1.00
0.50	3.4	1.3	0.92	1.00
0.60	2.4	0.40	0.970	1.00
0.70	1.8	0.12	0.990	1.00
0.80	1.27	0.035	0.996	1.00
0.90	0.60	0.009	0.999	1.00
1.00	0.006	0.000	1.000	1.00

Figure 21   Temperature, density, mass, and luminosity of a sunlike star in the center, at the surface, and at nine equidistant levels indicated by the radius.

corresponding to different times in the star's life. Today these same calculations are done in a few seconds by the modern electronic computers, which have provided astronomy with a new and powerful research tool. Thanks to the computer, we can now follow the star in the course of its evolution.

One starts by calculating the initial model of the star, which is assumed to have a certain mass and chemical composition; for example, it can be assumed to consist entirely of hydrogen, which to a first approximation may be regarded as true. In the model thus calculated the rate of energy production depends on the burning material, which in turn changes because, as we have seen, the "burning" that generates energy is nothing but the conversion of one element into another, in this case hydrogen into helium. In time, as the original nuclear fuel decreases and a new one forms, the chemical composition changes, the mechanism and rate of energy production change, and so do the internal conditions of the star. At this point one must calculate a new model of the star based on a different chemical composition. Unfortunately, the procedure is far more complicated than it seems. To give an example, as the chemical composition and the physical state of the various shells change, their opacity changes, and this in turn affects the flow of energy and the very conditions of the shells.

In any case, once every factor has been taken into account, if the problem and the computer program have been written correctly, the computer will furnish the solution. Thus by calculating a series of models at suitable intervals of time we can trace the past and future histories of the star and then compare the results with observations. This step follows from what I said before. Two of the four quantities obtained by computation are temperature and luminosity, which, for each shell, are given by the solutions of the equations. If we take the values of the temperature and luminosity computed for the outermost shell, which is the visible one, and plot them on the H-R diagram, we can compare them with those obtained from observations. In the final analysis we can trace on the diagram the path of a star of a certain initial mass and chemical composition. Also, we can find out at which points the star remains for longer or shorter periods, that is, which surface temperatures and luminosities the star holds for long periods and which for shorter ones.

LIFE OF THE STARS

By reconstructing the life of the stars with stellar models, we find that it depends on the mass.

As long as they are on the main sequence, where they stay the longest, stars generate energy by burning hydrogen in either the proton-proton reaction or the carbon cycle. These two mechanisms of energy production lead to stars of different internal structures, which also evolve in two different ways. The type of thermonuclear reaction that occurs in the stars depends on their masses. Let us consider the two cases in turn.

Stars of $M > 1.1 M_\odot$ Calculations show that when the mass is larger than 1.1 solar masses the interior of the star reaches a temperature of 16 million to 30 million degrees, which is sufficient to ignite the carbon cycle. At the end of this cycle, the carbon and nitrogen that were involved in the reaction are regenerated and hydrogen has turned into helium, except for a small fraction of it that has turned into energy. The higher the temperature, the faster hydrogen burns and the more energy is produced. Since the temperature increases toward the center of the star (figure 21), it follows that the larger the mass of the star, the smaller the region around the center where most of the energy is produced. For example, in a star of 10 solar masses half the energy is generated by a mass that is only 0.2 that of the sun and is located just around the center.

So much energy produced in so small a volume is not easily carried outward. In effect, not all of this energy manages to escape; part of it comes back, causing the interior of the star to bubble like the water boiling in a pot. Of course, the outer part of the very hot gas manages to flow upward, where the gas, though hot, is far less hot than it is below. These upper layers are transparent to radiation and energy can escape into space as light, x-rays, ultraviolet rays, etc.

In the more massive stars, then, heat propagates by convection in the central regions and by radiation in the outer layers. As we say, these stars have convective cores and radiative envelopes (figure 22a).

Stars of $M < 1.1M_\odot$ When the mass of a star is smaller than 1.1 solar masses the internal temperature is less than 16 million degrees and energy is produced through the proton-proton reaction, which depends on the temperature less than does the carbon cycle. In this case the central region of energy production is 15% of the stellar mass; since the star is smaller to begin with, this region is much larger than that of a more massive star. Hence the energy can easily escape from the core without hindrance. However, since the whole star is at a lower temperature, the outer layers are cooler and denser and impede the outward flow of radiation. To keep on flowing the energy from below must push these layers away, thereby causing the turbulent motion that can be seen in the photosphere of our sun, which is a star of this type. In sum, stars less massive than 1.1 solar masses have radiative cores and convective envelopes (figure 22b).

Finally, in very small stars (less massive than $0.2M_\odot$) energy propagates wholly, and everywhere, by convection (figure 22c).

Setting aside these pygmies for the moment, let us return to the heav-

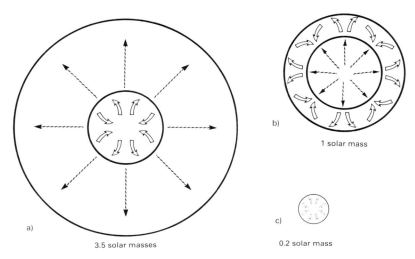

Figure 22   Sections of three stars showing how energy is propagated in their interiors depending on the mass: (a) convective core, radiative envelope; (b) radiative core, convective envelope; (c) transport by convection in the entire star.

ier stars. By violently burning hydrogen in the carbon cycle, the more massive stars produce an enormous amount of energy, which is necessary to support the great weight of the overlying layers and to keep the star in equilibrium. The lighter stars produce a smaller amount of energy but are equally capable of supporting the upper layers, which are less heavy. By burning hydrogen faster, however, the more massive stars have a much shorter life. Calculations show that a sun-type star burns hydrogen and remains on the main sequence for 10 billion years, whereas a star of $5M_\odot$ remains there for 68 million years, one of $15M_\odot$ for 10 million years, and one of $30M_\odot$ for only 5 million years. By showing that the most massive stars spend the least time on the main sequence, these results explain the lack of stars on the upper part of the sequence in the older clusters. It remains to be explained how the stars that leave the main sequence become red giants. The theory of stellar evolution has given us the answer.

THE RED GIANTS
When all the hydrogen in the hot core of a star has burned to helium the thermonuclear reactions come to a stop and so does the production

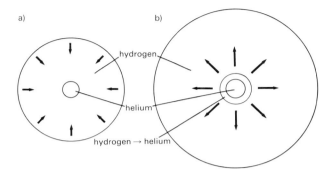

Figure 23   Schematic drawing showing how a main-sequence star evolves into a red giant. (a) After the formation of the helium core energy production comes to a stop and the outer layers, no longer supported by radiation pressure, fall toward the center; the gravitational energy thus developed turns into heat. (b) The temperature on the surface of the helium core rises enough to rekindle, in a thin shell around the core, the thermonuclear reactions that turn hydrogen into helium, generating energy. The newly produced helium augments the core, while the energy generated in the thin shell causes the overlying envelope of unburnt hydrogen to expand. The star, which still consists mostly of hydrogen, swells into a giant. The two sections are not to scale. For a detailed description of the interior of a red giant see appendix 1, illustrated with scale drawings.

of energy. No longer supported from below, the overlying material falls inward (figure 23a). Half the energy generated by the contraction tends to flow outward, while the other half heats up the layer of unburnt hydrogen surrounding the core, raising its temperature high enough for thermonuclear reactions to occur. Energy now is produced in a thin shell surrounding the core by the conversion of hydrogen into helium, as had happened before in the core. This energy pushes away the upper part of the star, which expands to great distances and cools down (figure 23b). A red giant is born—a star with a very small and dense helium core surrounded by a shell in which hydrogen is burning and by a distant envelope at a temperature of about 3,000°K, much lower than the temperature it had before expanding, which might have been even higher than 20,000°K. Between the hydrogen shell and the outer surface there is a gas so rarefied that if it were not for the great heat coming from the central region, a spaceship could venture through it.

While enormously larger than the main-sequence stars from which they originate, the red giants have the same or even a smaller mass. This makes sense since after all they are nothing but swollen main-sequence stars.

This is valid essentially for the outer envelope. Let us see now, instead, what may be happening in the interior. Before we go on, however, let me remind you that a star is a sphere of gas and behaves as such even in the central regions, where the pressure and density are very high. Hence the equation of state for perfect gases holds in its interior. According to this physical law, in a given volume of gas, say a cubic meter, the pressure is directly proportional to the temperature and the density, which, as everyone knows, is mass divided by volume. This law is the *deus ex machina* that keeps a star in equilibrium. Everything happens in a very simple way. When the energy generated by the thermonuclear reactions raises the temperature, then, in accordance with the law, the pressure also must rise; this increase causes the gas to expand and occupy a volume greater than it had originally, and the density of the material decreases. Consequently the temperature will fall, by the same law, and this will have the effect of slowing down the thermonuclear reactions. Thus as long as the interior of a star behaves like a perfect gas this mechanism acts as a thermostat keeping the nuclear reactions under control. In time, however, the interior of a star can become very much denser due to a modification in the structure of matter; this happens, for example, when the pressure from above, no longer balanced by the production of energy, keeps on compressing the material in the central regions.[10] Under these conditions matter is called "degenerate" and has very different properties from those of perfect gases. First of all it no longer transfers heat by convection or radiation, as normally happens in stars, but by conduction, as if it were (but it is not) a metal. Besides, and more important, it is much less compressible; the equation of state for perfect gases is no longer valid and temperature and pressure no longer depend on each other. Under these conditions, in other words, the temperature-pressure thermostat shuts off.

Now let us see what actually happens in the central regions of a red giant. All the hydrogen in the core has burned into helium. This in turn can burn into carbon and oxygen, both elements of greater atomic weight. But the nuclear reactions that permit this new transformation require temperatures of 100 million degrees. This is a much higher temperature than the 20 million degrees that made possible the conversion of hydrogen into helium. However, even this temperature can be reached by gravitational contraction; it all depends on the mass, which once again is the determining factor in the next step.

All the stars with masses exceeding $0.2M_\odot$ become giants,[11] but not all of them in the same way. In stars of $0.2-0.5M_\odot$ the gravitational contraction is not sufficient to raise the internal temperature to the point of igniting the helium, though it does reach the much lower temperature that permits the burning of hydrogen in the shell next to the core. The energy thus produced manages to inflate the outermost layer and the star becomes a giant.

In stars with masses between 0.5 and $3M_\odot$ there forms a degenerate helium core surrounded by a shell of hydrogen burning into helium and generating energy that supports the external envelope, more rarefied than the air we breathe. In this case, too, after the expansion of the outer layers, in the giant that has formed there remain the helium core and the overlying shell of hydrogen burning into helium. The helium forming in the shell complicates matters because it falls onto the core and increases its mass, thereby causing a continuous contraction of the core, which raises its temperature. Eventually the temperature reaches the 100 million degrees necessary to ignite the helium. With the thermostat shut off (because the matter is degenerate), this raises the temperature further. The increase in temperature, not being able to raise the pressure (again because the thermostat is not working), accelerates the thermonuclear reactions, thereby producing an enormous amount of heat, which in turn causes a sharp rise in the temperature itself. Hence the helium burns ever more rapidly, forming carbon and oxygen and producing more and more energy until, in a few seconds, as much energy is generated as all the stars in a galaxy normally produce. All this energy suddenly erupts in what astrophysicists call a "helium flash," a term that gives but a pale idea of a phenomenon that is truly inconceivable. Even though it is a very powerful flash, it would never blind anyone because, occurring inside the star, it is practically invisible. The energy thus developed reinflates the core that has produced it. The dilated core is no longer degenerate, the thermostat reactivates, and the remaining helium (about 95%) continues to burn in a stable way, forming carbon and oxygen.

Let us see now what happens to stars more massive than $3M_\odot$. In this case the interior of the star reaches the temperature necessary to burn the helium before the core degenerates. Thus the thermonuclear reactions gradually produce enough energy to support the upper layers of the star, which remains in equilibrium and burns helium into carbon and oxygen in a nonviolent way and without experiencing the helium flash. When all the helium in the new core is exhausted, except for a thin external layer, the nuclear reactions in the center stop; the external layer of unburnt helium contracts, developing heat and increasing its own temperature until it starts burning in a thin shell, as happened before with the hydrogen. At this point in the interior of the star there are two, nearly adjacent, shells in which nuclear fuel burns: an outer one of hydrogen that burns into helium and an inner one of helium that burns into carbon. Under both of them there is a core of carbon and oxygen whose future development again is determined by the mass.

If the mass of the star is less than $4-5M_\odot$, the core of degenerate carbon-oxygen never reaches the temperature necessary to ignite new nuclear reactions. For larger masses, but smaller than $9M_\odot$, when the inner temperature is on the order of 1 billion degrees, something similar to the helium flash occurs—a rapid igniting and burning of the carbon and oxygen, which may be violent enough to destroy the whole star in a colossal explosion. Finally, if the mass exceeds $9M_\odot$, the kindling temperature is reached inside the core before its gas degenerates. The star remains in equilibrium because energy is produced in its interior, and in a nonviolent way carbon and oxygen are transformed into neon, sodium, magnesium, silicon, and sulfur. At still higher temperatures, even these elements begin to undergo complex thermonuclear reactions, which in a very short time cause them to turn into iron. It is in this manner that the heavier elements are built from hydrogen in the more massive stars.

This process, however, does not lead to the formation of all the elements that exist in nature; while the fusion of atomic nuclei lighter than iron liberates energy, for elements heavier than iron fusion needs energy to occur. At this point, even though the star has a mass sufficient to build an iron core, the formation of the elements comes to an end. Its interior will be formed of several shells of different chemical compositions (figure 24).

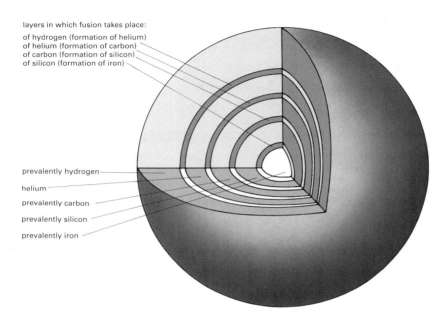

layers in which fusion takes place:
of hydrogen (formation of helium)
of helium (formation of carbon)
of carbon (formation of silicon)
of silicon (formation of iron)

prevalently hydrogen
helium
prevalently carbon
prevalently silicon
prevalently iron

Figure 24   Schematic drawing (not to scale) showing a massive red giant in the last stage of its evolution, shortly before exploding as a supernova. [From N. Henbest, *L'avventura dell'universo*, Laterza, 1980]

With this we have come to the end of the evolutionary path from the main sequence to the red giants and supergiants. The theoretical calculations that have enabled us to follow the evolution of stars of different masses furnish their surface temperatures and luminosities at different stages. These data are plotted in the H-R diagram shown in figure 25.[12]

If you examine figure 25 with the help of table 3 you will find the lifetimes for stars of different masses, starting from the zero-age line. In particular, notice that the more massive, and hence more luminous, stars take only thousands of years to travel the long tracks from 4 to 5 and from 3 to 6, whereas stars the size of the sun or a little larger take hundreds of millions of years to travel the much shorter tracks between the same points. This explains why in the H-R diagram for clusters shown in figure 20 there is a great empty space in the upper part between the main sequence and the red giants, while in the lower part, where stars have about the mass of the sun, this space is filled with a continuous band of stars—in the regions of the H-R diagram where

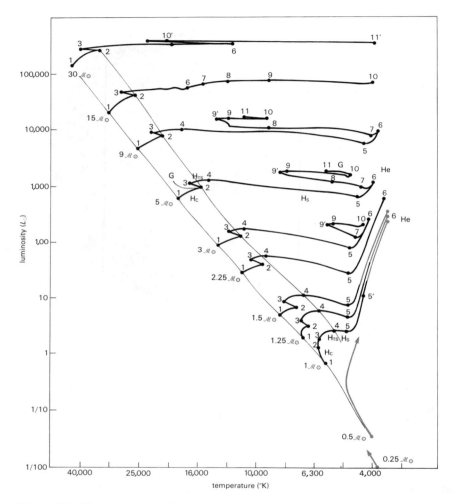

Figure 25   Evolutionary paths for stars of 0.5–30M$_\odot$ starting from the moment they are born, that is, upon reaching the zero-age main sequence (points 1). The regions indicated with letters correspond to H$_c$ = hydrogen burning near the center of the star; G = gravitational contraction of the entire star; H$_{TS}$ = hydrogen burning in a thick internal envelope; H$_S$ = hydrogen burning in a thin internal envelope; He = helium burning near the center and hydrogen burning in a thin shell. Table 3 lists the times needed for stars of 30–1M$_\odot$ to move from one numbered point to the next. [From R. Stothers, *Astrophys. J.* 143:91, © The University of Chicago Press]

Table 3   Intervals of time between successive phases (from 1 to 11') for stars with masses ranging from 30 to $1M_\odot$ [a]

Intervals between the positions of the stars on the evolutionary paths drawn in figure 25	$30M_\odot$	$15M_\odot$	$9M_\odot$	$5M_\odot$	$3M_\odot$	$2.25M_\odot$	$1.5M_\odot$	$1.25M_\odot$	$\dfrac{1.0M_\odot}{\text{(billions of years)}}$
	(thousands of years)		(thousands of years)		(millions of years)				
1–2	4,800	10,100	21,440	65,470	221.2	480.2	1,553	2,803	7
2–3	86.4	227	605.3	2,173	10.42	16.47	81	182.4	2
3–4	↑	↑	91.13	1,372	10.33	36.96	349	1,045	1.2
4–5	~10	75.5	147.7	753.2	4.505	13.10	104.9	146.3	0.157
5–6	↓	↓	65.52	485.7	4.238	38.29	≥200	≥400	≥1
6–7	↑	717	490	6,050	25.1	—	—	—	—
7–8	53.1	620	95	1,020	40.8	—	—	—	—
8–9	—	190	3,280	9,000	—	—	—	—	—
9–10'	↓	35	155	930	6	—	—	—	—
10'–11'	1.3	—	28.6	76.9	—	—	—	—	—

a. Adapted from I. Iben Jr., *Ann. Rev. Astron. Astrophys.* 5:571. The path for stars of $30M_\odot$ was published by R. Stothers, *Astrophys. J.* 143:91. The zero-age line, that is, the time of the "birth" of the star, corresponds to points 1. Notice how rapidly massive stars cross the regions 3–6, 4–5, and 6–11. This is the reason why the corresponding part of the H-R diagram appears almost entirely devoid of stars.

the stars remain longer it is easier to find them. If we go at random to an office that is open only an hour a day, chances are we shall find it closed, but a place open continuously for eight hours seems to be open all the time.

THE END

As we have seen, the evolution of a star cannot proceed beyond the formation of an iron core. In the case of the heaviest stars this phase marks the definitive arrest of the production of nuclear energy inside. It is the end. No longer supported by the energy produced in the central regions, the material quickly collapses inward, freeing an enormous amount of gravitational energy that tears the star apart in a colossal explosion, pulverizes it, and throws the remains into space in all directions. A part of this energy disperses the material, but a part of it serves to initiate the fusion of elements of atomic weight lower than iron, which turn into elements of higher atomic weight. Thus while the star is violently destroying itself, in a very short time it forms most of the elements heavier than iron, like silver, gold, mercury, and lead. When this tremendous catastrophe is visible from the earth it seems as if a star has suddenly burst into brilliance. We call it a supernova.

As we have seen, it is only the very massive stars that reach the supernova stage. But stars of smaller mass, from $3-4M_\odot$ to $9M_\odot$, also can explode when their degenerate carbon-oxygen core suddenly ignites. In these circumstances, too, elements heavier than iron are built, such as nickel, copper, and zinc. These different types of explosions occur in stars of different chemical composition and mass and may account for the different types of supernovae.[13]

But not all the stars terminate with an explosion. Next we shall see how others end and what is left of stars that, though capable of shining for millions or even billions of years, are equally destined to come to an end like everything else.

BLACK DWARFS AND WHITE DWARFS
We shall start with the smaller and lighter stars.

$M < 0.01M_\odot$ When the mass is less than a hundredth that of the sun, the initial collapse and gravitational contraction are not sufficient to heat the interior of the star to the point of igniting thermonuclear reactions. Energy is produced only by gravitational contraction and is transported through the interior exclusively by convection, until the matter degenerates. A star of this type lives only hundreds of millions of years, shining dimly with a reddish color due to its low surface temperature. Gradually, as it grows cold, the star darkens and disappears from view. It has become a black dwarf and its life ends even before it begins, never having reached the main sequence.

$0.01M_\odot < M < 0.085M_\odot$ In stars of masses ranging from a hundredth to almost a tenth the solar mass, the gravitational contraction is sufficient to raise the internal temperature to around a million degrees, which makes it possible for heavy hydrogen (deuterium) to burn to light helium. Thus these stars reach a sort of main sequence and emit light, though faintly, until they exhaust the deuterium, degenerate, and end up as black dwarfs like the stars of smaller mass. The duration of their life also is on the order of hundreds of millions of years.

Black dwarfs have never been observed, but seven stars are known to have invisible companions that might be black dwarfs; although the companions cannot be seen, they were detected from the perturbations in the motions of the stars. On the other hand, it is believed that an enormous number of them might exist, and their presence would solve, at least in part, the mystery of the invisible masses detected even in the vicinity of the sun.[14]

$0.085M_\odot < M < 0.2M_\odot$ If you recall, in stars with masses between nearly 0.1 and 0.2 that of the sun, the heat is transported by convection throughout the entire star. Consequently, the hydrogen continuously mixes, turns entirely into helium, and there is no hydrogen left to form the shell where by burning into helium it would generate the energy needed to raise the outer envelope. On the other hand, when all of the star's hydrogen has turned into helium as a result of the thorough mixing, there is no longer an outer envelope to raise. Thus these stars never become giants. On the contrary, they contract slightly, and the internal material becomes a degenerate gas of pure helium. Nevertheless, they do not burn out rapidly like the red dwarfs. Because of its very high internal temperature, estimated to be over 10 million degrees, the star emits a great deal of light and heat through an outer layer of nondegenerate material that is almost like a film and very hot, generally between 5,000 and 20,000 degrees. Due to this high temperature the star looks white. These strange stars that never swell into giants but contract and degenerate are very small objects. It has been calculated that their radii cannot exceed 14,000 kilometers, which is a little more than twice the earth's radius. For these reasons they have been called "white dwarfs." [15]

These stars keep on shining by dissipating the heat drawn from their interiors, and not having any mechanism to produce more heat, little by little they deplete it. Thus also for them the moment arrives when the surface temperature begins to decrease, and eventually they redden and expire. But it takes a very long time for this to happen. The more massive white dwarfs, of about $0.2M_\odot$, are estimated to live around 2,000 billion years, while those of $0.085M_\odot$ may live twice as long. Since the universe began about 15 billion years ago, white dwarfs born at that time not only are not extinct but can be considered in early youth. Their lives are so long that we do not know whether the universe itself will last that long—so long that they practically merge into eternity. But it is not eternity, and these stars, like all the others, will come to an end. $0.2M_\odot < M < 3M_\odot$ Stars between 0.2 and 3 solar masses evolve in different ways, as we have seen, but their end is always the same: a white dwarf that slowly but inexorably burns itself out.[16] If the mass is smaller than $0.5M_\odot$, it will be a white dwarf composed entirely of helium; if the mass is larger than $0.5M_\odot$, it will be made of carbon and oxygen. In this second case, before ending as a white dwarf the star will turn into a strange object that until a few years ago we did not know how to interpret.

THE PLANETARY NEBULAE
Toward the end of the eighteenth century the great astronomer W. Herschel discovered small, strange nebulae that he called "planetary nebulae" because of their compact and round appearance, which made them resemble the planets. Their existence and how they fit into the general picture of the universe remained a mystery until recently, but meanwhile investigations of these objects produced a wealth of information. Their gas was analyzed and turned out to be like that of the great nebulae, ionized, and at a temperature of 10,000°K. It was found that the disk is as large as or a little larger than the solar system and

that the total mass is a few tenths that of the sun.[17] It also was discovered that at the center of each planetary nebula there is a star with a surface temperature that may reach 100,000 degrees, a luminosity 10,000 times that of the sun, and a very small radius, just a few tenths that of the sun. Such peculiar characteristics make the nucleus of a planetary nebula very similar to a white dwarf. The similarity has been confirmed by another fact, namely, the velocity with which the two types of stars move inside the Galaxy. Taking all the planetary nebulae within a radius of 5,000 light-years, their mean velocity turns out to be 38 km/sec, almost the same as that of the white dwarfs, which is 41 km/sec.

If the white dwarfs represent the last stage in the evolution of small- and medium-sized red giants, the planetary nebulae, whose nuclei are very similar to white dwarfs, may represent the transition phase. This hypothesis is supported by the fact that the gaseous envelope of a planetary nebula appears to be expanding outward from the central star. We know that the material in a star's core can degenerate, which is just what happens in white dwarfs. We also know that the shell of burning hydrogen and possibly that of helium cause the upper layers to expand. If some mechanism keeps these layers expanding they can reach the size of a planetary nebula and become so diluted, so transparent, that the central degenerate core will be seen through. The fact that the envelope is expanding has another implication. As it gets increasingly larger and more rarefied, at a certain point the gas envelope must become so tenuous that it will disappear. As a matter of fact, planetary nebulae vary in size, and the largest are also the faintest, with a few exceptions—notably, the well-known nebula in Aquarius, which appears larger than usual only because it is much closer to us. The gas envelope is believed to last about 30,000 years.

Systematic all-sky surveys have led to the discovery of more than 2,000 planetary nebulae. Considering that we see only the nearest ones, and assuming that their distribution is fairly uniform, we find that at present there must be about 40,000 of them in the entire Galaxy. It has been calculated that 2 to 3 planetary nebulae must be born every year in our Galaxy alone.

Thus it should not be too difficult to find one that is just forming and to see if the process occurs as I have described it on the basis of conjecture more than solid evidence. And perhaps we are finding them. In the past twenty years astronomers have observed a dozen or so strange objects that may very well be planetary nebulae in the process of forming.[18]

Take, for example, the star V 1016 Cygni, which in 1947 appeared to be a red giant with its spectrum of late M type. Between 1963 and 1968 it increased in brightness a hundred times and then displayed a spectrum similar to that of a planetary nebula. In addition, it is surrounded by envelopes that seem to have formed in different epochs. In the past few years this star has been the object of intense study, particularly at the Asiago Observatory, and the body of evidence indicates that it is a red giant turning into a planetary nebula under our very eyes. Tens of thousands of years from now the envelope perhaps will be completely dissolved and only the nucleus will be left—a white dwarf. In the context of the initial analogy—where we compared the study of stellar evolution to the problem of discovering the life span of an unknown population in a very short time—this discovery would mean that we have found an individual with brief but significant signs of a serious illness that also will determine the rest of his life and the type of death that awaits him.

The skies are not immutable as people once thought, and this we have known for quite some time. But even today there is a belief that evolutionary changes, however rapid, are so slow compared with human life that we cannot possibly observe any of them. But now we see that this is not true. Logic, observation, and the aid of other sciences are enabling us to obtain an ever more complete and detailed picture of how the universe evolved in the billions of years when we were not there and how it will evolve after we have ceased to exist. Our assiduous scrutiny of the sky has been rewarded by the feeling that we are personally witnessing a rare phenomenon, a particular point in the great evolutionary cycle of a star, something that we never thought we could achieve in the brief course of our lives.

THE SUPERNOVAE

We have seen that, almost certainly, stars of small or medium mass turn from red giants into white dwarfs through the planetary-nebula phase. Theory had already told us so because, after all, inside the red giant there is already a white dwarf (the degenerate core). But now we know several stars that appear to be evolving under our very eyes, and we are observing them carefully in the hope of obtaining conclusive evidence. But this is not the only evolutionary phenomenon that we can actually observe, for we are constantly witnessing another phase in the life cycle of the stars—their death, or at least the death of stars that end in a certain way.

Death comes quickly to human beings, and it seems even quicker when it is due to a sudden violent accident. As it happens, there are stars that end this way. Every year we see at least ten stars that die a violent death through an explosion that tears them to pieces, like the unlucky man who has stepped on a mine.

This is what happens to the more massive stars when their nuclei, whether made of carbon or heavier elements, suddenly ignite and almost instantaneously liberate an enormous amount of energy. It also can happen when, the production of energy having stopped, the outer layers collapse and, all of a sudden, acquire a tremendous kinetic energy at the expense of the gravitational potential energy they possessed while they were supported by the pressure from below. All this energy serves to build the heavier elements, to disperse the outer layers into space, and to compress the inner material to the point that it becomes denser than degenerate matter. This is the case of a supernova explosion. Its effects are tremendous and at the same time extraordinary, and its importance will manifest itself again and again and in an ever more pervasive way as we continue our journey in time. Right now we shall follow it at the moment in which it happens, when the event is at its most impressive if we witness it from nearby. In the course of the explosion the star ejects a great deal of material and radiation that might not just dissipate harmlessly into space and disappear. Both can destroy celestial objects nearby—the star's own planets, for example (if any)—and kill life forms as far away as the planets of neighboring stars. But while the matter and energy that the supernova hurls into space may have disastrous effects, the energy it directs inward has strange and wondrous consequences, which in large part (but not entirely) are the fruit of theoretical studies.

Neutron Stars   As F. Zwicky had predicted as far back as 1933, the collapse of a star can compress its matter into a body with a diameter of only about 30 kilometers. In such a body the free electrons penetrate the atomic nuclei and by neutralizing the positive charge of the protons form neutrons. The latter are so close together that it is as if the star had become a single, enormous atomic nucleus.

In the white dwarfs, even though matter is concentrated, there still remains some empty space; in the neutron stars even this remaining space is filled up. Thus neutron stars are not only much smaller than white dwarfs but also much denser. On the earth, a cubic centimeter of one of the most massive white dwarfs would weigh about 80 tons, but a cubic centimeter of a neutron star would weigh between 100,000 and 1 billion tons.

There is a fundamental difference between the discovery of the white dwarfs and that of the neutron stars. The white dwarfs were observed in the sky and subsequently astronomers and physicists, struck by their strangeness, developed the theory. The neutron stars, instead, were predicted and calculated as a consequence of supernova explosions without having ever been observed. I might add that their smallness makes it almost impossible to observe them in any case. Astronomers have discovered objects such as the pulsars and various phenomena that appear to substantiate their existence; nevertheless, many things remain to be explained.[19]

Black Holes   The collapse and supernova explosion do not always form a neutron star. Once again it all depends on the mass. It has been calculated that a neutron star cannot have a mass larger than about $3M_\odot$. Of course, the exploding star can have a larger mass because part of its material is ejected into space and does not contribute to the formation of the central object. But if the collapsing material has a mass in excess of $3M_\odot$, the collapse never stops and the star tends to be reduced schematically to a point. According to the theory of general relativity, this high concentration of mass in a very small space causes such a pronounced curvature of the space itself that it becomes a sort of funnel capable of swallowing anything that comes within its gravitational field; not even light can escape from this trap, which consequently appears to be black. This funnel—the notorious "black hole"—may or may not have a bottom. Perhaps, just as it has an entry, it also has an exit, although not necessarily in our own space-time. Thus if we were to enter a black hole, we might emerge into another space-time, that is, another universe; or we might come out in another time and another place in our own universe; or we might come out an instant later in a place very far away from the entrance, or in the very place from which we exited but in another time, either in the remote past or the distant future.

These and other wonders of the black holes open up fantastic horizons, including the hope of practical applications.[20] But here we had better stop. Not because the black holes are unimportant in the economy of the universe. Recent elaborations of the theory show that space might be full of small black holes (of mass slightly or a great deal smaller than the sun's) that were born with the universe. It is also believed that there might be huge black holes at the center of some globular clusters and many galaxies, including our own, that formed at a later time or might still be in the process of forming and accreting. There might be collisions between black holes and other celestial bodies and, finally, in a distant future, even a general collapse of the whole universe into a single, immense black hole that would swallow up every existing thing in a vortex with no escape and no end. All of this would be not just important but revolutionary in our study of cosmic evolution, as we shall see later. It bears reminding, however, that the existence of black holes has not yet been proved. Furthermore, their existence, which is required by theory to explain the end of massive stars, has begun to appear a little less necessary.

As we have seen, the formation of a black hole in a supernova explosion depends on the mass of the collapsing material. We have long known that there are stars 20 or 30 times more massive than the sun, that is, much heavier than the $3M_\odot$ limit above which the collapse cannot be halted and must result in the formation of a black hole. This is why in theory black holes should exist.

But the life of a star, such as we have described it, may be affected by other events. In particular, there is a phenomenon that complicates matters quite a bit and might even demonstrate that, no matter how large its initial mass, a star need not end as a black hole. I am referring to the loss of mass, to which we shall now turn our attention.

COMPLEX STARS

Thus far in our exploration of stellar evolution we have proceeded on the tacit assumption that the stars were single and that their mass always remained the same (except for the small fraction that turns into energy). In reality, at least half of the stars are double or multiple systems, and furthermore, there is increasing evidence that in the course of their lives many stars lose a part of their mass. Of course, this does not invalidate what has been said so far. First of all, our conclusions are still valid for single stars and for all those that lose a negligible amount of mass, and in any case, they are the starting point for understanding a stellar evolution complicated by these new conditions. Studies of the more complex stars have begun in earnest only in recent years, and consequently no comprehensive theories have yet been developed. Here we shall briefly examine a few cases just to demonstrate that the stars are far more complicated than our theories had envisaged them, not only in their characteristics but also in their transformations in time.

THE LOSS OF MASS

We already know one way in which the stars can lose part of their mass to space—through nova and supernova explosions; in most cases, after a time, the expelled envelope becomes clearly visible. But there also are noncatastrophic cases in which a star loses a part of its material in a continuous and quiescent manner. The fact is that a star is not delimited by a well-defined surface: its outermost layers merge gradually into space in clouds or veils of gas and dust. The farther away they are from the center of the star, the less the force of gravity affects them, and other forces, originating essentially from the energy produced in the interior, tend to blow them away.

Take the sun, for example. It has been calculated that through the solar wind it loses at least $10^{-14}M_\odot$ every year. This means that every million years it loses a hundred-millionth of its total mass. Not a great deal. But the sun is a small star, and the gas in its atmosphere is not as rarefied as the gas in a red giant's atmosphere, which is more than 10 million times less dense, extends to distances from the center 500 to 1,000 times greater, and is held by a gravitational force 25,000 to 1 million times weaker. As a result, a red giant loses from 10^{-8} to $10^{-6}M_\odot$ per year, which in a million years means a hundredth of the sun's mass to 1 solar mass. Although most of these results have been obtained spectroscopically, sometimes the material lost to space can be photographed, particularly when it has moved far enough from the star not

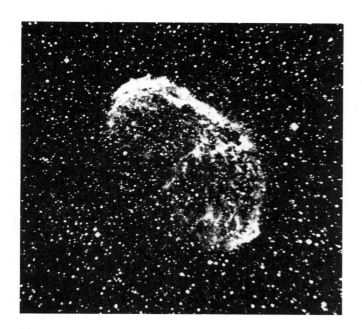

Figure 26   Material ejected by a star in a late evolutionary phase. This nebula is known as NGC 6888. The star is not easy to identify because it is located behind the central filament. This photograph was obtained by the author with the 60-centimeter telescope of the Loiano Observatory (Bologna).

to be hidden by its overwhelming light (figure 26). Of course, the loss of mass might not continue at this rate for a very long time.

The loss of mass also occurs in very young stars, and it is in this field that the most significant discoveries have been made. First of all, there is the phenomenon of the so-called Be stars. Their broadened and complex emission lines have revealed that these objects have very high rotational velocities, as high as 500 km/sec, in contrast to 4 km/sec for stars like the sun. Because of their very rapid rotation they become flattened and develop a disk or ring in the equatorial plane, from which a part of the material escapes into space. The mass lost every year by this celestial spinning wheel can go from a ten-millionth to a billionth of the mass of the sun. As we discussed earlier, young stars, or rather protostars, also can lose mass through the phenomenon called "fuor." But the most spectacular case of mass loss is that of the very young, very massive, and highly luminous stars located on the upper part of the main sequence, that is, O stars and B stars of the first spectral subdivisions. This discovery, due to the use of artificial satellites above the earth's atmosphere, is fairly recent and came as quite a surprise.

As you know, O stars are the hottest (30,000°K) and the most luminous (100,000 times more than the sun). Since they emit mostly in the ultraviolet, their study has seen considerable progress only in the past decades, thanks to the observations performed from space by scientific satellites. Note that the ultraviolet radiation is emitted by the outer layers of the stellar atmosphere, while what we perceive as light emerges from deeper layers. Ultraviolet observations of these stars have revealed a P Cygni effect,[21] which means that they are surrounded by envelopes expanding at velocities of 1,000–2,000 km/sec. Through studies of the entire electromagnetic spectrum (visible, infrared, and radio) astronomers have been able to calculate the amount of mass that leaves the star. The first complete analysis, performed for the star ζ Puppis, revealed a loss of 7 solar masses every million years. Every O star with a surface temperature higher than 30,000°K shows a P Cygni profile in the lines of oxygen ionized 5 times and nitrogen ionized 4 times. This means that the stellar wind in the outer layers has a temperature of more than 100,000°K; hence the temperature is increasing outward. The physical mechanism that causes this increase is not known, but we are not concerned with this fact at the moment. The important point is that for a time a part of the star's material is blown away. The question is, How long does a star like ζ Puppis lose mass? If we assume that the initial mass is $60-100M_\odot$ and that the loss of mass occurs during the entire time in which hydrogen burns in the core, and considering that the length of this period in turn depends on the mass, we find that it should last 3 million years, with a total mass loss of $21M_\odot$.

The evolutionary models for stars that lose mass have been developed only recently, and one of the first discoveries is that the loss of mass slows down a star's evolution and extends the period of hydrogen burning in the core.

Mass losses of this magnitude have been discovered in stars other than ζ Puppis, and while this research has only just begun, one thing is already evident: The stars that lose mass at a tremendous pace are the most luminous, which, as we know, are also the heaviest. Thus there seems to be a little more justice in the sky than here on earth;

those that have more, pay more. For the stars, on the other hand, this fact could be a piece of luck. Thanks to the loss of mass they could reach the last stage—the collapse—with their masses reduced below the fateful limit at which the collapse cannot be stopped. In other words, through the loss of mass they would never become black holes. If this conclusion could be generalized, the existence of black holes would no longer be necessary to explain the end of massive stars.

But apart from this, the discovery of large mass losses is very important because it changes, if not the substance, at least the pace of stellar evolution, especially for the more massive stars.

Matters become even more complicated when the mass that escapes from one star falls onto another one nearby, something that can happen in binary systems.

DOUBLE STARS
A binary system, which consists of two stars revolving around the common center of mass, can be of two types—wide and close. In the first case the components are separated by distances up to 1,000–2,000 times the earth's distance from the sun, which is 149.6 million kilometers and is called an astronomical unit (AU); in a close binary system the two stars are so close that their surfaces can even touch.[22] In a wide binary system the two orbits, which may be more or less elliptical, are traveled slowly in tens or hundreds of years; in a close binary system the orbits are nearly circular, naturally much smaller, and the stars travel them at dizzying speeds in a matter of days or even hours.

Double stars are something of a mystery from the time of their birth. A single star, as we have seen, originates from a cloud of gas and dust that starts contracting, collapses, and then ignites. Three different theories have been proposed for the formation of binary systems, but the right one seems to be connected to the theory of the origin of single stars.

According to one of the theories, a binary system would form from the chance encounter of two stars that come close enough to capture each other and remain physically bound by each other's attraction. Such close encounters are extremely unlikely, however, due to the vastness of space. The rarity of encounters is in total disagreement with the large number of existing double stars. Clearly binary systems do not form in this manner, or do so only exceptionally.

Another theory affirms that double stars would form from the breaking up of stars rotating at high velocities, but this theory, too, does not seem to agree with observations. Actually, there are stars that spin very rapidly, such as the Be stars we saw earlier. In Be stars, however, the high rotational velocity produces a ring or disk but does not break the star in two. Furthermore, such a mechanism would explain the formation of close binary systems but not of wide ones. Thus this theory is not wholly acceptable, even though one cannot exclude the possibility that it might work in some cases.

The third theory was sketched by Laplace almost two centuries ago and is still the most general and, paradoxically, the most modern. In its most convincing form, this theory is based on an idea suggested by C. M. Varsavsky in 1960 and developed qualitatively by P. Giannone and M. A. Giannuzzi. It starts with a dense, dark cloud of material such as a Bok's globule. This is quite reasonable considering that some of them have masses tens or hundreds of times that of the sun and consequently should produce whole groups of stars including, in all likelihood, double or multiple systems. Confining himself to the simplest case, which, however, does not exclude the others, Varsavsky proposed that a cloud in gravitational contraction may split in two and form two distinct nuclei separated by distances ranging from a few fractions of to a few hundred astronomical units. The orbit of each condensation may have any eccentricity whatever. If it is greater than or equal to 1, that is, the orbit is hyperbolic or parabolic, the two stars separate and move away from each other forever, becoming single stars and evolving as such. If, instead, the eccentricity is less than 1, the two protostars travel elliptical orbits, are physically bound to each other, revolve around the common center of mass, and keep on moving through the residue of the material from which they were formed. This material has the effect of slowing them down, and little by little, their orbits tend to become circular. The closer the nuclei are at the start, the more efficient this process will be. Let us consider the two extreme cases by changing only the initial distance between the stars. If that distance is great, the orbits are wide, the stars travel them slowly, and the scattered material dissipates before it has a chance to slow the stars down, draw them closer together, and make their orbits circular. In this case, the orbits may be highly elliptical, the stars remain far apart, and we have a wide binary system. If, instead, the initial distance between the nuclei was small,

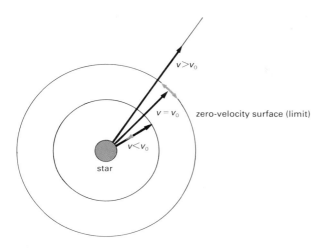

Figure 27   A particle leaving a single star with a velocity v lower than, equal to, or greater than the escape velocity v_0 will, respectively, fall back on the star, remain indefinitely at a given distance, or escape into space. The velocity vectors are drawn on a plane containing the center of the star.

due to the attrition with the scattered material over a large number of revolutions, the stars draw closer together and move faster, their orbits become circular, and we have a close binary system.

This theory, which has the advantage of having a single mechanism explain both types of binary systems, is supported by the discovery of protostars that are already double and often widely separated; according to the other two theories this could not occur, but it is quite normal if we assume that the two stars originated from the same cloud.

Now that the binary system is born, let us see how it evolves. In the wide systems the two stars usually are so far apart that they do not influence each other, and hence evolve like normal single stars. In the close systems it is an entirely different story. To understand what happens we must stop a moment on some results of mathematical physics that most of humanity is happily unaware of but that once understood will open a window on very strange worlds whose existence we had never suspected.

Suppose we have a single spherical star (figure 27) and suppose further that a particle leaves the surface of the star, moving outward along a radius, with a certain velocity v. Due to the star's gravitational attraction the particle will gradually slow down, and at a certain point it will stop and fall back onto the star. Since the star is spherical, all the particles that leave with the same velocity will come to a stop at the same distance before falling back. This distance defines a spherical surface (obviously ideal) where the particles leaving from all points with the same velocity have an instant of pause. For this reason it is called the zero-velocity surface. Naturally, there are as many zero-velocity surfaces, all concentric, as there are velocities that the particles can assume.

But there is a limit to all things. In our case, the limit is the escape velocity, which we shall denote v_0. The escape velocity is that value of the velocity that cannot be exceeded if one wishes an object (a stone or a particle) launched from the surface of a celestial body to fall back on it instead of escaping into space forever. The escape velocity naturally depends on the attractive force of the body from which the stone or the particle are launched, that is, on the body's mass and on the distance between its center and the person who is launching it. For the earth, at sea level, it is 11.2 km/sec; for the moon, of smaller mass, it is 2.38 km/sec; and for the much more massive sun, it is 617.7 km/sec.

Getting back to our star, a particle that leaves its surface with a velocity equal to the escape velocity will never fall back, but it will be able to move in any direction over the outermost zero-velocity surface. All the particles that leave the star at precisely the escape velocity reach this surface and remain there, moving freely in any direction, while the particles departing with higher velocities will go past it and disperse into space.

This is a very simple case. Things become more complicated when there are two stars in close proximity, but we do not have to worry about that; the calculation of the corresponding surfaces was done a long time ago, and we shall turn directly to the results.

Figure 28 shows the intersections of the orbital plane of the stars with the two most significant surfaces, denoted by the letters A and B. The star on the right (primary) is larger and more massive than the star on the left (secondary). Because of the combined effects of two attractions, the zero-velocity surfaces (A) of the stars are not two spheres but two sorts of pears known as Roche lobes. The point L_1 at which the two Roche lobes are joined is called the "inner Lagrange point." If a particle travels with a velocity lower than the escape velocity for surface A, denoted v_{0a}, it remains within the lobe containing the star. If its

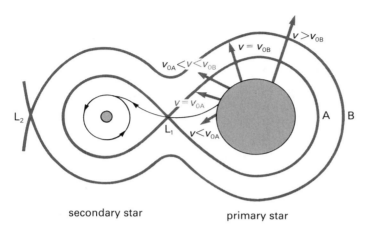

Figure 28   Schematic representation of a close binary system showing the zero-velocity surfaces and the Lagrange points L_1 and L_2.

velocity is equal to v_{0a}, the particle reaches surface A, whereupon it belongs to both stars and moves around both of them without ever leaving A. If it travels with a velocity between v_{0a} and v_{0b}, it reaches an intermediate region where it belongs to the system but to neither star in particular. Finally, if its velocity is equal to the velocity v_{0b} necessary to reach surface B, it moves along this surface until it reaches the outer Lagrange point, L_2, through which it escapes into space. Obviously, any object that reaches B with a velocity higher than v_{0b} will escape from the system. You might think that all possibilities have been exhausted, but there is another very important case, namely, that of a particle that sets out from the primary in particular directions with a velocity higher than v_{0a}. In such a case, after crossing the point L_1 the particle is captured by the secondary and may either fall directly upon it or enter into orbit around it. If the latter happens for a large number of particles, a ring will form about the secondary.

Now that we have considered the various possibilities schematically, let us see what happens with real stars, with actual masses, radii, and distances, such as we observe in the sky. The basic configurations of close binary systems, as shown diagrammatically by Z. Kopal in the early 1950s, may be reduced to eight. These configurations are very interesting, not only because they describe strange objects that actually exist, but also, and above all, because they do not always represent permanent situations for various types of double stars; often, they represent different phases that the same binary system goes through in the course of a process of continuous transformations.

Figure 29 illustrates all eight cases. The sections of the Roche lobes and the Lagrange point L_1 are drawn on the assumption that the primary star, of larger mass, is on the left-hand side and that the secondary star on the right has a mass two-thirds that of the primary.

Case (a) shows a wide binary system whose components do not interact or exchange mass and evolve independently of each other. In (b) the more massive star is assumed to have an envelope that fills its Roche lobe; if part of this material moves with a velocity higher than v_{0a}, it crosses over to the other Roche lobe through L_1 and may form a ring around the less massive star. Case (c) is the exact opposite of the previous one. In (d) both stars have "light" envelopes in which the particles move with velocities not high enough to let them escape from the stars, but sufficiently high to let them travel from one star to the other. Case (e) occurs when the more massive star has reached the phase of a red giant losing mass to the secondary; a typical example is the star UX Monocerotis. Case (f) is the opposite of the previous one and corresponds to a red-giant secondary losing material to the primary; well-known examples are U Cephei and U Sagittae. Finally, there are cases, such as (g), in which both stars fill their Roche lobes, and others, such as (h), in which the expansion of the stars has proceeded to the point that they take on the look of a dumbbell. In the last two cases the stars are in contact, the exchange of mass is extensive, and part of the material is also dispersed around the system, forming rings or spirals that swirl around the double star, fade with distance, and dissolve into space. This case, especially if we assume that phase (h) comes before (g), appears to be a beautiful example of the formation of a double star from the breakup of a single star. In 1975 V. G. Gorbatsky did in fact suggest that stars of the W Ursae Majoris type may form in this manner, but his theory has run counter to actual observations. Once again, this demonstrates how careful one must be in formulating theories even if they are perfectly logical and mathematically sound.[23]

EVOLUTION OF BINARY SYSTEMS
We have made hypotheses regarding the origin of double stars; we have taken a sample of binary systems and discussed them in some de-

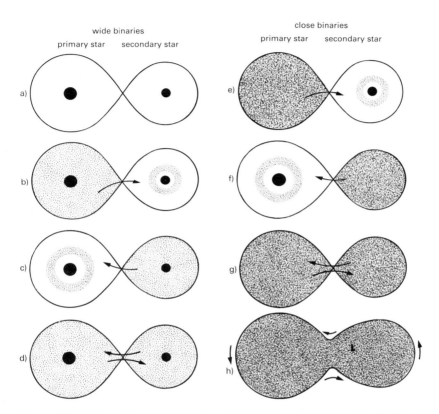

Figure 29   Schematic representation of eight binary systems after Z. Kopal. In each system the secondary (right) is 2/3 the mass of the primary (left). [From *Sky and Telescope*]

tail; we have seen that in the close systems matter can flow from one star to the other and examined the routes it travels, but we still do not know how these stars evolve. Although it is evident that their evolution must be different from that of single stars, we do not yet know what the differences are. There is nothing strange in that. What was true for single stars—namely, that they do not all follow the same evolutionary path—is even more true for close binary systems, since the presence of each star modifies the life of the other. Now that we have met them in all their variety and have acquainted ourselves with the methods used by theoretical astrophysicists to explain the behavior of both single and double stars, we can attempt to reconstruct their lives. Out of the many possible cases let us choose one that has been extensively studied.

Without bothering to redo the astronomers' calculations, but on the basis of the results they obtained, we shall look for a double star with specific characteristics and shall get close enough to it to observe it well. Actually, at such a short distance we would not be able to remain alive through the transformations of the stars, but we shall approach them only in thought. We must abandon reality in another sense also: For us, time will speed up, so that we can see in a few minutes what actually occurs over millions of years.

Let us approach a close binary system just formed and composed of a star of 1 solar mass and another 10 times heavier. As usual, we shall call "primary" the more massive star and "secondary" the other one, and we shall continue to use these designations even though, due to their evolutionary peculiarities, the secondary might become more massive than the primary. The two stars have formed at the same time, and since for a while each evolves in the region delimited by its Roche lobe, they both live their separate lives as if they were single. But the more massive star evolves much more rapidly than its companion and in a relatively short time becomes a red giant, while the secondary remains practically unchanged. While turning into a red giant, the primary first fills and then expands beyond its Roche lobe. As we look on in wonderment, the star becomes a sort of huge red pear from whose tip (point L_1) matter begins to escape and to fall toward the secondary. The latter, initially less massive, little by little acquires more mass by taking it away from the primary, and at a certain moment begins to swell, becoming huge and red. What has happened? It is not difficult to explain. The matter escaping from the giant has markedly increased the mass of the secondary, and by the evolutionary process we have already seen (which occurs faster, the larger the mass) this star has evolved into a red giant in a much shorter time.

The exchange of mass between the two bodies changes their periods of revolution (by Kepler's third law) and the stars move closer and closer. When their masses become the same they reach the minimum distance from each other and revolve around the common center of mass with the maximum velocity. At this point we behold a truly paradoxical event: the primary, initially more massive but now lighter, keeps on losing material to the secondary, now heavier, and cannot help doing so as long as it is large enough to spill out of its Roche lobe.

But let us not lose sight of the secondary. By capturing mass from its companion, it has accelerated its own evolution and at a certain moment it expands to the point of filling its own Roche lobe. Now matter flows back from the secondary to the primary, which reacquires, at least in part, the mass it had lost. But while the secondary has been growing in mass and volume at the primary's expense, the latter has been declining for a long time; it has gone through the red-giant phase and turned into a white dwarf.[24] The material escaping from the secondary is now falling on this last remnant of the massive star that evolved too fast. We now see a white dwarf surrounded by a tenuous gas ring, or disk, that revolves around it and is continuously replenished and augmented by material coming from the red, dimly shining companion. Having filled its Roche lobe, the latter keeps on losing material to the primary through the escape point, L_1. Although reduced to a small white dwarf, the primary attracts this material, and the more so the closer it comes, so that the in-falling material continues to accelerate until it collides with the gas disk. The impact heats up the disk, forming on it a hotter, brighter region, commonly called the "hot spot" (figure 30).

The incoming material not only forms the hot spot but gradually increases the mass of the disk, generating an imbalance that progresses to the point of causing an explosion. After equilibrium is reestablished the white dwarf lives on quietly until the material that keeps on coming from the secondary causes another imbalance and a new explosion.

At this point we shall leave our observatory in space and return to the earth to see if among our papers and observations we can find something that recalls the strange events we have observed during the

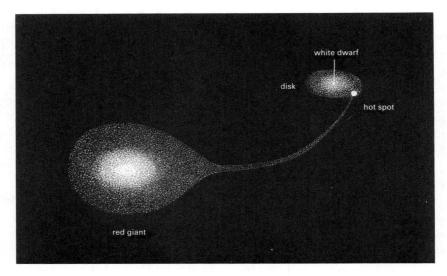

Figure 30   Artist's conception of a nova going to the explosive phase. The material escaping from the red giant forms an accretion disk around the white dwarf. Observe the "hot spot," the hotter and brighter region where the material falls.

last stage in the evolution of our double star. It is not hard to find. The explosive phenomena we have witnessed from space are the outbursts of novae, recurring novae, and the so-called dwarf novae. Observations, especially of the last ones, have shown that although the star undergoing the outburst appears to be single through the telescope, it is in fact composed of two very close stars, a white dwarf and a red subgiant. Through research initiated in the 1950s and brought to a high degree of precision by means of spectroscopy and superfast photometry, we also have found evidence of the disk, the jet of material that falls on it, and the hot spot.

These phenomena have been studied most extensively in the case of dwarf novae, where they are most frequent, particularly by B. Warner and his associates. But novae also appear to behave in this manner. The general consensus today is that all novae are recurrent; that is, they all go through a number of outbursts. The only difference between one nova and another would be the length of the interval between two successive explosions and the extent of the explosion. The rarer the phenomenon, the bigger the explosion, and hence the brighter the flash of light. The different types of novae would be due to differences in the stars that make up the binary systems. According to G. T. Bath and G. Shaviv novae and recurrent novae correspond to two different types of close binary systems. In the first case the white dwarf has as a companion a star still on the main sequence, material accretes more slowly to the disk around the white dwarf, and the explosion occurs approximately every 100,000 years and therefore appears unique to us whose lives are so much shorter. In the second case the companion is a red giant that loses mass to the disk much more rapidly, the explosion occurs every 30 years, and the nova appears to be "recurrent." In the dwarf novae the outburst is smaller but even more frequent, occurring at intervals of a few dozen days.

Thus the study of the evolution of close binary systems has led us to the solution of the mystery of the novae that so impressed the ancients, who were shaken in their belief of the immutability of the sky by the sight of a star that seemed to appear and disappear under their very eyes. Today we know better. The novae are not at all new stars. On the contrary, they are bodies that have precociously aged because of the exchange of mass, and now and then experience climactic periods that reveal their slow agony.

All the changes we observed earlier in the evolution of one particular binary system correspond, in a certain sense, to the evolutionary tracks of single stars, and can generate, depending on the initial masses, a great variety of systems in which the components' masses, distances, velocities, shapes, colors, and diameters change with the passage of time in the most disparate ways from one system to the other.

There is the case of the wide binary systems whose components evolve without interacting, as if they were single stars. And the case of the W Ursae Majoris stars, in which the two components are joined forming a single, strange body that rotates furiously about itself (figures 29g and 29h). Then there are a number of cases in which the primary is so massive that after a rapid evolution it self-destructs in a supernova explosion of such violence that the system is disrupted and the secondary flies off into space and keeps moving away from the cataclysm at a crazy speed as if it were terrified by the end of its companion.[25]

Neither calculations nor fantasy can depict for us *all* of the actual cases of double stars, in different stages, with the most disparate ways of evolving, that populate the heavens and pass and repass over our heads every night by the thousands, by the millions. Yes, observation has revealed to us many of them and theory has predicted many others. But it is certain that in the sky there must be an endless number of cases still undiscovered, just as among our papers there are perhaps others that calculations show to be perfect in every detail but that in reality do not exist and may never exist.

A CYCLICAL PROCESS?

Our first exploration of the stellar world in time is over. Later on we shall embark on a far more ambitious journey that will take us from the remotest past to the distant future. Meanwhile, by observing a great variety of stars and with the help of theoretical studies, we have learned many things: how the stars are formed; how they live while remaining for a shorter or longer time on the main sequence; how they evolve into red giants; how they expire as red or white dwarfs; why some end in a supernova explosion; what the novae are and how they form. But above all we have discovered two fundamental facts: one, that stars are continually forming and ending; and two, that they are formed from dust and turn back into dust.

Have we finally found a cyclical process? It would seem so. The stars are born from gas and dust and end in gas and dust from which new stars are born, and so on for ever and ever, always ending and always beginning again. A moment's reflection, however, will tell us that things do not go exactly this way. The material from which the stars originate is processed in their interiors and partly transformed into heavy elements, and it is this new material that is eventually redistributed in interstellar space, perhaps in very large amounts. Since it is richer in heavy elements, this material is different from the primordial matter and forms stars that are not perfectly identical to those from which it originally came. Furthermore, the interstellar medium changes not only in quality but also in quantity because a good part of the mass of dead stars is left in those small but superdense cosmic relics we call white dwarfs and neutron stars. Hence this cycle, too, is an illusion. In time, unless unknown phenomena intervene, the interstellar matter will change and diminish to the point that no new stars will form.

Let us not worry now about that faraway time. Instead, let us retrace our steps. We have seen that a star is born, lives, and dies. We have talked a good deal about its beginning and its end, but not enough about what may happen in the course of its life while it is settled on the main sequence for a period that, for light stars like the sun, lasts many billions of years. Given its stability, there may not be much more that we can discover about the star itself. But in its vicinity there may be something less stable and far more interesting, something that formed with the star or from the star and that is worth exploring in time even more than the star: a system of diverse bodies orbiting the star, bodies on which unfolds the extraordinary process of life—we ourselves.

3 FORMATION OF THE PLANETARY SYSTEMS

ORIGIN OF THE SOLAR SYSTEM

Among the many stars that formed, are forming, and will form, we now shall follow only one. A little over 4.6 billion years ago a vast interstellar cloud of gas and dust began to collapse under the pull of its own gravity, heating up little by little from the conversion of gravitational energy into heat. Gradually it shaped itself into a roundish globule that continued to shrink and get warmer. After a time the globule broke up into fragments, and going back to that remote age we shall now follow the fate of one of those fragments, which we shall stop to observe from a suitable spot in space, neither too near nor too far.

The fragment has a strange appearance. It does not look like a piece of solid body, as the word would suggest, but rather like a dark, undefined cloud with irregular contours, and so tenuous that only from a distance can it be seen as a dark stain on the starry background, a large region of which it obscures. The contracting cloud has reached the diameter of the present solar system, but continues to shrink and get hotter. As it contracts, the cloud begins to glow, first a barely visible deep red, next red, and then orange. Finally, when it has shrunk to the size of the earth's orbit, its surface temperature reaches 4,000°K. At this point the cloud begins to emit a great deal of energy in the form of yellow-orange light. Our dark cloud has grown smaller and rounder and, above all, has started to shine; it has become a star, or rather a protostar, which one day will be our sun.

THE SUN IN ITS BIRTH
The surface temperature is not too high and the light emitted by the unit of surface area—say, a square centimeter—is a little weaker than the light now emitted by a square centimeter of the solar surface. But being so large, the protostar has a surface area more than 10,000 times greater than that of today's sun; hence its luminosity is almost 10,000 times greater than our sun's. This intense orange light illuminates a flattened but very extended nebula within which there are discontinuities that indicate a lazy revolution about the center. Inside the nebula the earth also is forming; our world is in the making.

But let us continue to follow the protosun, which at this point looks very luminous because of its size but intrinsically is not very hot or luminous. In its interior, in fact, the temperature is still low, so low that the nuclear reactions cannot start. Energy continues to be produced by gravitational contraction, and the protosun emits half this energy inward and half outward as radiation. Its luminosity decreases because the surface area shrinks, and the surface temperature increases so little that it does not compensate for the shrinking of the radiating area. For approximately 10,000 years the luminosity of the protosun is about 100 times greater than that of today's sun. Gradually the energy directed inward raises the temperature of the central region to 1 million degrees and hydrogen begins to burn slowly, partly contributing to the luminosity. The contraction continues, causing the internal temperature to rise toward 10 million degrees. The fusion of hydrogen with production of helium increases rapidly until, the internal temperature having reached 10 million degrees, the fusion of hydrogen in the proton-proton reaction produces all the energy necessary to keep the sun in equilibrium. The contraction stops; the star has reached the zero-age line; and the sun is born.

According to the latest calculations this stage has lasted 30 million years. From the moment of its birth the sun remains on the main sequence, undergoing only small, very slow variations. The central temperature again rises until it reaches the current value of about 15 million degrees. In the sun's interior this increase in temperature starts hydrogen burning also through the carbon-nitrogen cycle. (This reaction produces only 10% of the energy; the remaining 90% comes from the proton-proton reaction.) Its effect on the exterior is a 50% increase in the sun's luminosity.

FORMATION OF THE SOLAR SYSTEM

While the sun was forming, the vast disk of gas and dust surrounding it also was changing. We last saw it, flattened and very large, in the strong hot light of the orange protosun. Now we want to learn more about it: what it was made of; why it had flattened; and how it later formed planets, satellites, meteorites, and comets—in sum, all the bodies that today constitute the solar system.

First of all it has been conclusively established that the disk originated from the same material that formed the sun, that is, from the same cloud, from the same fragment of globule that by collapsing gave rise to the protosun. In fact, some elements, such as deuterium, lithium, and iron, are found in the same abundance in interstellar matter as in the planets and meteorites. In the sun there is the same percentage of these elements, except that deuterium and lithium are reduced to 1/100; this is not surprising considering that in the stars these elements are rapidly destroyed in the first nuclear reactions. Furthermore, some meteorites (ancient residues of the original material) contain isotopes resulting from the radioactive decay of heavy elements produced only in supernovae. Thus it has been deduced that the cloud that formed the solar system had been enriched by material coming from at least two supernovae. The second explosion may even have caused the formation of the solar system by starting the contraction of the parent cloud.

The original solar nebula contained gas (prevalently hydrogen and helium) mixed with a very small percentage of heavy elements such as carbon, silicon, and iron, joined in particles no heavier than a thousand-billionth of a gram. During the gravitational contraction of the nebula these particles collided and, helped by electrostatic and electromagnetic forces, often combined into larger particles. The entire cloud of gas and dust should have converged toward the center from all sides. However, this did not occur because the cloud also was in rotary motion about an axis. Due to this motion, the inner part of the nebula and the material distributed along the rotation axis, or along directions slightly inclined to it, collapsed into the protosun. In contrast, the force of gravity that attracted the remaining material toward the center was balanced by the centrifugal force due to the rotary motion; the closer the material was to the plane perpendicular to the rotation axis and the farther it was from the center, the more efficient was this balance of forces. Under the pull of gravity, the particles, which meanwhile had combined into larger grains of masses from 1- to 10-billionths of a gram, settled on the equatorial plane and formed a thin disk wrapped in a much thicker gas disk (see figure 31).

The nebula circling the protosun extended more or less as far as the present planetary system, and its mass cannot have been much larger than the total mass of the bodies that comprise it today. Had its mass been larger, a second star, perhaps more than one, would have formed instead of the planets, and the planetary system would have belonged to a double or multiple star. Since this did not happen, the mass of the protoplanetary nebula cannot have been much more than 1/100 that of

Figure 31   Illustration of the forces at play in the primitive nebula from which the sun and planets are about to form. The pull of gravity (black arrows) causes the material (especially the dust) to fall toward the equatorial plane and the protosun. The centrifugal force (white arrows) acts counter to the fall of the material toward the protosun. Pressure (gray arrows) counteracts the fall of the gas toward the equatorial plane, and stopping it gives a marked thickness to the gaseous component of the disk. [Adapted from Field, Verschuur, and Ponnamperuma, *Cosmic Evolution,* Boston: Houghton Mifflin]

the sun. Hence the density of the nebula must have been very low— more or less the density of a vacuum obtained in our laboratories.

When the sun, still on its way to the zero-age line, had a luminosity about the same as today, the temperature at the distance of today's earth must have been around 0°C. At this temperature the solid grains consisted of metal aggregates, silicates, and water ice. By now, in this region of the future solar system, and naturally in the entire region closer to the sun, there no longer were hydrogen, helium, and noble gases because the solar wind had blown them to the regions where to-day we find Jupiter, Saturn, and the other distant planets. At those distances the temperature had to be below $-200°C$ and all the volatile compounds, such as water, carbon dioxide, methane, and ammonia, were frozen in crystals that mingled with grains of metals and silicates. But nowhere in the nebula was the temperature ever low enough to solidify hydrogen and helium, whose freezing points are very near absolute zero. Thus the protoplanetary nebula took the shape of a large doughnut filled with a thin layer of chocolate (figure 32). The dough-nut was made of hydrogen and helium, while the chocolate layer corre-sponded to the thin disk of solid grains. In the region near the sun, the nebula was almost nonexistent and the particles consisted of metals, silicates, and water ice; in the region of Jupiter and Saturn, hydrogen and helium gases were abundantly distributed in a fairly thick layer, while the solid particles in the thin disk also consisted of ice crystals of methane, ammonia, and carbon dioxide.

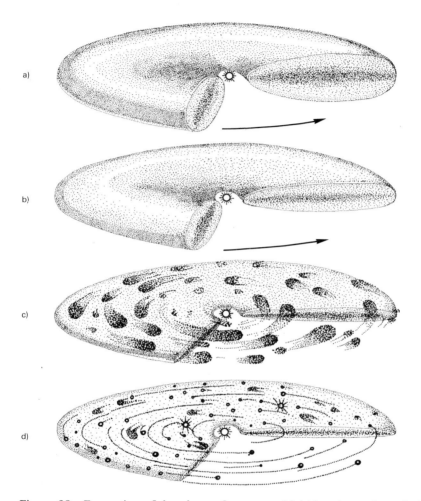

Figure 32 Formation of the planets from a toroidal (doughnut-shaped) cloud of gas with a thin layer of dust: (a) the first toroidal distribution resulting from the flattening of the cosmic cloud that has formed the sun (center) due to the effect of rotation—in this phase the separation of gas and dust has just begun; (b) the toroid has flattened further and all the dust has settled on the equatorial plane; (c) formation of the first swarms composed of particles already fairly large; (d) the planetesimals grow larger but fewer in number and are on their way to forming the embryos of the future planets.

All this material revolved around the sun, not as a rigid disk, but according to Kepler's laws, as today. In other words, every particle, no matter how small, behaved like a miniature planet. By Kepler's third law, the particles closer to the sun moved faster than those farther away from it. But, given two particles in adjacent orbits, the particle in the outer orbit was only a little slower than the one in the inner orbit. Thus, no matter how fast they moved around the sun, their relative velocity was very low, so low that if one day their orbits brought them together, they would not collide violently and break up into smaller pieces, but would approach each other slowly and combine gently into a single larger and heavier particle.

Given the enormous number of particles and adjacent orbits, larger particles did form, some massive enough to attract and capture the smaller ones in their vicinity. Thus, little by little, some particles increased in mass and volume at the expense of their smaller neighbors. These larger bodies have been called planetesimals. However, the above mechanism was not the only one responsible for the formation of planetesimals, nor initially was it the most important one. At that time the velocities with which particles of that size were randomly moving were higher than the relative velocities required for a process of direct accretion such as I have described. To explain the first steps in the formation of planetesimals V. S. Safronov and, independently, P. Goldreich and X. R. Ward have had recourse to collective mechanisms. Such mechanisms may be visualized as the fragmentation of a very thin dust ring embedded in gas. In this manner one obtains a large number of bodies, on the average 1 kilometer across.

As it traveled along its orbit, every planetesimal would sweep the surrounding space, absorbing the smaller bodies and thereby growing in mass and capturing still more bodies. In this way a planetesimal ended by forming around the sun a circular band relatively clear of matter. Meanwhile, the planetesimals began to collide, sometimes shattering, and sometimes, if the impact had not been violent, joining and forming still bigger bodies. Regardless of the number of destructive and constructive collisions, it is clear that, if the latter occurred, in the long run the process was constructive because the fragments from the destructive collisions sooner or later would be captured by larger planetesimals. In this manner, with the passage of time there formed larger bodies that we call embryos because they were the "embryos" of the present planets.

The embryos evolved into planets by different processes. For Mercury, Venus, the earth, and Mars it was a process of accretion by the mechanism that we have seen for planetesimals. When each of the four embryos had swept its band of space around the sun, collecting all the minor bodies that had moved into its sphere of action,[26] the process of accretion came to an end. The four planets formed from particles of metals, silicates, and ice and were nearly devoid of the primeval gases that at the time of the accretion had already been blown away to the more distant regions of the solar system.

In the case of Jupiter and Saturn, the solid bodies, having added ice crystals of different composition, acquired masses 2–3 times that of the earth and in the process reached a critical mass that enabled them to start attracting the hydrogen and helium abundantly distributed in the surrounding space. As each planet grew in mass by capturing these gases, its sphere of action broadened and new material was collected. According to Safronov and E. L. Ruskol, the formation of the solid nuclei took from 10 to 100 million years, and the gas accretion from 100,000 to 1 million years. Finally, when the two bands of the protoplanetary disk were entirely emptied of material the process came to an end.

The cases of Uranus and Neptune are again different. These two planets formed much more slowly than the others. In the first place, the planetesimals took much longer to sweep up their regions due to their slow motion around the sun; in the second place, there was much less gas available at those distances since meanwhile a great part of the hydrogen and helium either had been captured by the nuclei of Jupiter and Saturn (quicker to sweep their orbits) or had left the solar system and dissipated into space while Uranus and Neptune were going about their lazy orbits. Nonetheless, Uranus and Neptune managed to collect a certain, not very large, amount of hydrogen and helium. As a result, they are not as rich in these two elements as they are in ice.

The fact that Uranus and Neptune were slow in sweeping their orbits favored the formation and survival of minor bodies composed essentially of the same material as these planets, that is to say, comets. It has been calculated that by the action of the great planets, which are known to be powerful manipulators of cometary orbits, many of the newly formed bodies were flung far out into space, forming what J. Oort envisaged as a comet reservoir.[27] Once in a while one of them leaves its distant abode to return to its birthplace, bringing us a planetesimal or an embryo that 4.5 billion years ago might have evolved into a planet but today instead is only a luminous, evanescent streak across the sky. Recaptured by the sun on a smaller orbit, this body, which now appears in the familiar shape of a comet, repeatedly loses the material from which it formed to the space that once was its cradle. Remember, comet, that you were dust and to dust you shall return.

No need to bemoan the inevitable end of all that is born. We shall see later on that this is not the whole story and that not even the life-death cycle is a real cycle because nothing that ends (and everything seems destined to end) dies completely, and nothing and nobody leaves the world the same as it was before.

Furthermore, it is not the time to think of death when we are witnessing a birth, the birth of the solar system—a system that would have such a development that for thousands of years we would believe it was our whole universe, a system that we have not had time to know well even in its current aspect and, above all, that has an ever richer future ahead that we shall never know.

THEORIES AND OBSERVATIONS

My description of the birth of the solar system is based essentially on the theory by V. S. Safronov and the line followed at the Institute of Earth Physics in Moscow, where for over forty years they have been studying theories and results from all over the world in an effort to reach a simple and comprehensive cosmogonic theory. But are we quite sure that the solar system was formed in this manner? So that you may better assess the validity of this theory, which is the most acceptable to date, at least in its general lines, I shall put it in its historical context along with other theories and present some supporting evidence for it.

VARIOUS HYPOTHESES

Any hypothesis about the origin of our planetary system must start from some fundamental characteristics of both the system and the individual planets that can be deduced from observations. Needless to say, this is done as a matter of course by every scientist who formulates a theory. In my previous description these characteristics either were taken for granted or were explicitly mentioned. However, before we continue, it might be a good idea to recall some basic facts.

First of all it should be noted that the orbital planes of the planets are only slightly inclined on the mean plane of the entire system. The motion of all the planets occurs in the same direction, which is also the direction in which the sun rotates about its axis and the satellites revolve around their respective planets. There are very few exceptions to this rule—the comets, a few asteroids and satellites, and the planet Pluto, which is perhaps an exsatellite of Neptune. The orbits of the planets are elliptical, as dictated by Kepler's first law, but so little elongated that they look almost circular. The mean distances of the planets from the sun follow a strange empirical law, known as the Titius-Bode law, for reasons that we cannot explain. Finally, as we move away from the sun we find, first, small, very dense, and rocky planets—Mercury, Venus, the earth, Mars—and the swarm of asteroids; then, gigantic but not very dense planets—Jupiter, Saturn, Uranus, and Neptune—which are believed not to have a solid crust; and finally Pluto, another small planet, which, as I said, seems to be an exception.

Theories about the formation of the solar system have been advanced for the past two and a half centuries. Basically, they all fall into two categories—one that is catastrophic, and one that I shall call nebular. Heading the first group is the theory of the French naturalist G. Buffon who about 1745 suggested that the solar system might have formed from material ejected from the sun as a result of the fall of a comet. The first theory of the nebular group was enunciated by the philosopher Kant in 1755. According to Kant, the sun and the planets were born from one flattened nebula; the more massive particles within the nebula attracted the lighter ones in their vicinity and, by accretion, formed the planets. The satellites were born in the same manner.

In 1796 P. S. Laplace formulated another nebular theory that subsequently was combined with the previous one, so that today one usually speaks of the Kant-Laplace theory. In effect, they are two distinct theories. Kant's hypothesis is the first sketch of the current planetesimal theory. Laplace, instead, starts from a cloud made entirely of gas and arrives at the formation of rings that by shattering give rise to the planets. The nebular theory as formulated by Laplace had some weak points. Nevertheless, it was accepted in its entirety throughout the nineteenth century, perhaps because of Laplace's renown in the field of celestial mechanics and the rigor, as well as the elegance, of his formulation. Toward the end of the century, however, J. C. Maxwell criticized the theory with such solid arguments that the scientists' faith in it began to be shaken. As a result the beginning of our century saw the reappearance of catastrophic theories, even though in a form different from the rather naive one proposed by Buffon.

In essence, though with some variations, these theories maintained that the planetary system was born from the sun's encounter with another star that passed close enough to cause enormous tidal phenomena. The passage of the perturbing star would have sucked from the sun a large, cigar-shaped jet of material whose thicker middle part gave rise to the major planets, Jupiter and Saturn, and the two ends to the minor planets.

If this theory were true, the solar system would be an exception in the entire Galaxy because the distances between stars are very large and there is a very small chance that two stars should pass close enough to cause tidal phenomena of sufficient magnitude to form planets. But there are other difficulties. The encounter theory does not explain some of the characteristics of the solar system, such as the circularity of the orbits and the Titius-Bode law. Finally, the very hot material drawn from the sun does not seem at all suitable for forming planets, and on the other hand, as L. Spitzer calculated, it would have dispersed into space before cooling off enough to form them.

Since the 1940s the nebular theories have returned to the forefront, and today we have a large body of evidence to support them, especially those based on planetesimals. The photographs of planets and satellites taken by various planetary probes, particularly in the 1970s, have shown that all the bodies unaffected by erosion, that is, bodies with no atmosphere (Mercury, the moon, some of Jupiter's and Saturn's satellites) or with a very thin atmosphere (Mars), are full of craters (figure 33). Most of these craters are very ancient. They are nothing but the scars left by the fall of smaller bodies near the end of the accretion period, when planets and satellites had already formed a solid crust but smaller bodies were still around in huge numbers.

A second piece of evidence is the scarcity of noble gases on the minor planets. The other gases were retained in the solid state in the interior

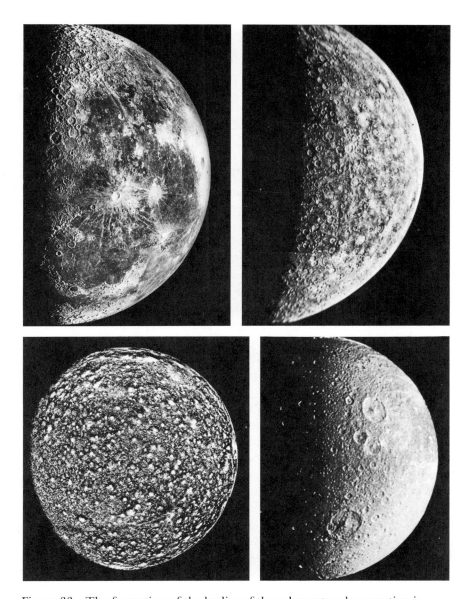

Figure 33   The formation of the bodies of the solar system by accretion is
proved in all the cases in which the traces, that is, the craters formed by the fall
of minor bodies, have not been subsequently erased by other mechanisms, such
as atmospheric agents. The four examples shown here are (top) the moon and
Mercury and (below) Callisto and Dione, satellites of Jupiter and Saturn, respectively. The phenomenon has also been observed on smaller bodies, such as
Phobos and Deimos (figures 41 and 42). [Courtesy Mount Wilson Observatory,
Las Campanas Observatory, Carnegie Institute of Washington, and NASA]

of the forming planet, as crystals and compounds, and then released when the temperature increased. Think, for example, of the hydrogen and oxygen incorporated as ice and later reemitted by volcanoes as water vapor. Noble gases such as helium, neon, argon, and xenon continued to be part of the protoplanetary gas and were dissipated into space with it, except for part of the more abundant helium, which, as we have seen, was captured mainly by Jupiter and Saturn.

A third confirmation of the theory was provided by a computer simulation of the formation of a planetary system. The first type of evidence is the most obvious; the second is more indirect and sophisticated; but this third one is certainly the most brilliant—as simple and amusing as a game. The game consists in trying to construct the solar system with a few reliable pieces, setting in motion the mechanism of the planetesimals, which we assumed to be the basis of the accretion theory, and then waiting to see what happens. This computer simulation was done for the first time by S. H. Dole in 1970. Here is how it works and what Dole found.

The computer is programmed with the assumption of a central star of the mass of the sun surrounded by a cloud of gas and dust in which the particles are in motion, all in the same direction and in elliptical orbits with the center of mass in one of the foci. The particles are confined within a sort of disk, that is, a particular solid called an exocone (figure 34). Within the exocone the ratio of gas to dust is constant and the density of both decreases regularly with the distance from the star. Into this cloud are inserted, one at a time, nuclei of condensation in motion on elliptical orbits. The dimensions of the nuclei and the orbital characteristics are chosen at random. As the nuclei grow by sweeping their orbits, annular lacunae form—if the bodies pass a certain critical mass they collect gas from the primordial cloud. If a new planet is brought by its orbit into interaction with a previously formed one, or if their orbits cross, the two bodies combine to form a single, heavier planet that continues to grow. In sum, starting from the actual conditions of the protoplanetary cloud, the computer takes us through, albeit schematically, the same steps that the accretion theory had formulated from real data for the real solar system.

The nuclei are inserted until all the dust has disappeared from the system. At this point the computer shows the masses and orbital parameters of the planets that have formed, and you cannot believe your

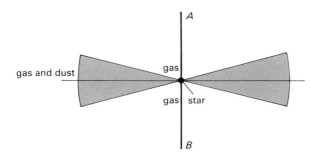

Figure 34   Section of an exocone, a sort of dish with the top and bottom surfaces converging toward the center (here occupied by the star). To visualize it
in three dimensions one must rotate the figure about the vertical axis AB. It is
a further schematization of figure 32a. [From R. Isaacman and C. Sagan,
Icarus 31:512]

eyes. The computer has drawn a number of planetary systems (figure
35). They are not all alike because nuclei and orbits were chosen at
random, but they all look like brothers of the real solar system, which is
mixed in with the simulated ones. It is in figure 35 and it can be easily
identified, though nobody can do it at first glance. See how long it takes
you to find it, covering the caption with one hand, and then try this test
on your friends.

    Let me point out some common characteristics. There is evidence
of regularities that suggest laws such as Bode's; the inner planets turn
out to be small and rocky; the intermediate ones are large and gaseous; and the outer ones are small. Although schematic, the mechanism really seems to work.

    In 1977 R. Isaacman and C. Sagan used Dole's program to construct
models of planetary systems with more varied initial conditions. They
used different laws of vertical and radial density distributions within
the primordial exocone and different laws of gas-to-dust ratios, and
they varied the total mass of the cloud and the eccentricity of the orbits
of the condensation nuclei. Thereby they obtained a great variety of
possible planetary systems, from systems of multiple stars accompanied
by planets, to systems with Jupiter-type planets at distances of hundreds of astronomical units, to planetary systems composed solely of
asteroids. In all models, however, the earthlike planets never became
more massive than 5 times the earth's mass. In addition, all the simu-

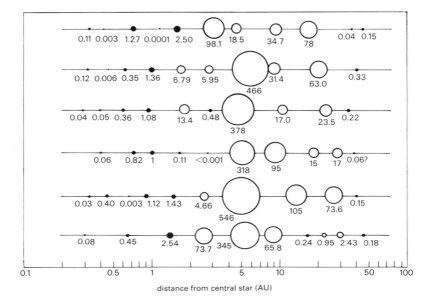

Figure 35   Some of the planetary systems obtained by Dole using a computer. The diameters of the planets have been drawn so as to be proportional to the cube roots of their masses, indicated by the numbers below each disk, with the mass of the earth as the unit of measure. The distances from the central star (in astronomical units) are marked on a nonlinear scale with the intervals growing shorter from left to right so that the inner planets do not bunch up and the outer planets do not move out of the picture, as would have occurred with a linear scale. Observe the similarity of the various planetary systems. The solar system is the fourth from the top. [From R. Isaacman and C. Sagan, *Icarus* 31:511]

lated planetary systems have two points in common with our own: the number of planets almost always is around 10 and some type of Titius-Bode law is found in every system, although it may be different from one system to another.

Besides all the evidence for the accretion theory, it may be interesting to mention a critical review made in 1976 by Tai Wen-sai and Chen Dao-han. They examined and discussed 40 theories concerning the origin of the solar system with special emphasis on two fundamental problems, namely, the source of the planetary material and the mechanism of planet formation. They came to the following conclusions. The planetary material was neither ejected nor captured by the sun and the entire solar system originated from one nebula. Planets and satellites developed from the nebular disk that formed around the sun, but the planets formed neither through rings, huge protoplanets, intermediate bodies, regular models, nor turbulent vortices. At first dust and ice particles combined to form planetesimals; then the latter combined to form planets. Dust and ice particles initially fell on the equatorial plane, and when the density increased the process of planet formation markedly accelerated. Finally, the rotation of the planets did not start from the tidal action of the sun on huge protoplanets, but rather as a result of the impact of planetesimals upon planetary embryos.

In conclusion, the planetary theory sketched by Kant more than two centuries ago seems to be the right one after all. And the view of the formation of our planetary system that we obtained by observing the cloud containing the infant sun should closely reflect what actually occurred around the sun about 4.6 billion years ago.

OTHER PLANETARY SYSTEMS

The fact that the accretion theory gives a satisfactory explanation for the formation and evolution of the solar system does not necessarily mean that a planetary system like ours will form for every star that is born; nor does it mean that if other planetary systems exist, they all formed in the same manner. Some might have formed by one of the mechanisms that we have rejected for our own system. But the nebular theory appears to be the closest to reality, and a planetary system is most likely to originate from the outer part of the cloud that forms a star. Thus one wonders whether it may be possible to discover planetary systems in the process of forming just as we have been able to find stars in the process of forming. After all, we know where to look for them—around protostars. Research in this field already has proved successful.

In 1976, while studying the spectra of T Tauri stars, A. E. Rydgren, and S. E. and K. M. Strom found that these protostars are surrounded by clouds almost as vast as the solar system and at temperatures close to $-100°C$ that contain silicates in the form of very fine grains of dust. Moreover, in recent years it has been found that several stars are surrounded by a gas disk. Within the gas disk almost certainly there also is a more or less distinct dust ring. From the point of view of the nebular theory the latter is the more interesting because the formation of planets is supposed to start with aggregates of dust particles.

In this context there are interesting results from infrared observations of young stars. Many of these stars show an excess of infrared emission, which means that more energy is emitted in the infrared than would be anticipated from the spectral distribution of the energy as determined by the temperature and deduced from the visible spectrum. The infrared excess is easily explained if we assume that there is a dust envelope that absorbs part of the light or ultraviolet radiation emitted by the star, warms up, and reemits the absorbed energy as infrared radiation. Since we see the star and the surrounding envelope as a single point source because of the distance, our instruments detect, superposed, both emissions—the weakened visible and infrared emissions; they also tell us which is the more intense of the two.

By observing a great number of these young objects it has been found that they fall into four categories:

1. objects that radiate only in the infrared;
2. objects that radiate prevalently in the infrared but also a little in the visible;
3. objects that radiate in the visible with a slight infrared excess; and
4. objects that radiate only in the visible.

The spectral distribution of the observed emission is illustrated in figure 36a. Quite likely, the four cases listed above constitute, in that order, an evolutionary sequence. But there are two possible sequences, as shown in (b) and (c). Both start from a spherical envelope of gas and dust that surrounds the star and prevents us from seeing it. At this stage all that reaches us is the infrared radiation emitted by the envelope. In the second stage, the envelope has contracted and the dust has separated from the gas. Up to this point, evolution has occurred in the same manner; but from here on it can take two routes. Either planetesimals and planetary embryos form from the dust while it gradually thins out and dissipates, or the disk disperses without forming any planetesimals or planets. In both cases, however, the energy distribution curves look the same, as shown in (a). Yet there is a fundamental difference, since in one case a planetary system has formed while in the

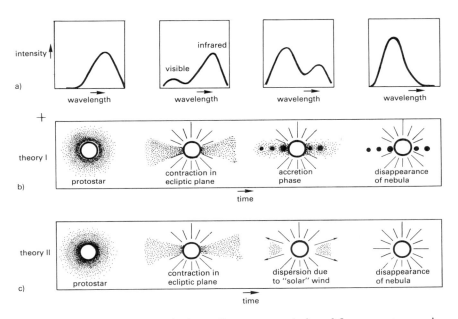

Figure 36   Two possible evolutions of a protostar deduced from spectroscopic observations. (a) Four curves show the distribution of the energy emitted by protostars; in each graph intensity is plotted along the ordinate and wavelength along the abscissa (with infrared on the right). (b) and (c) illustrate the stellar models that explain the graphs above. Observe that the last two graphs apply both in the case of a planetary system and in the case of a star that gets rid of the last traces of the dust envelope without having formed planets. [From W. K. Hartmann, *L'Astronomie* 94 (September 1980)]

other it has not. Which of the two evolutionary sequences should one accept? Both, I think.

The evolution of the dust ring can go either way depending on the type of star in its center. If it is a highly luminous star, of spectral types O and B, the surrounding gas and dust will be rapidly blown away, as demonstrated from observations of these stars in stellar associations. In this case we have sequence (c), in which the object quickly reaches the fourth stage, where the infrared excess disappears. If, instead, the star is of spectral type G, like the sun, or later (K–M), the dust settles in a disk because of the rotary motion and remains to form planets through the planetesimal mechanism, while the gaseous component of the envelope is slowly pushed away toward the outer regions. At the end, when al. or nearly all the dust has been transformed into planets, the space around the star becomes perfectly transparent, and in this case too the infrared excess disappears.

The conclusion that only stars of the later spectral types are likely to have planetary systems had already been reached by O. Struve many decades ago. And the most amazing and remarkable thing is that he arrived at it by an entirely different route. Struve's starting point was a fact that previously had been an obstacle to the acceptance of the nebular theory. The distribution of angular momentum[28] in the solar system is very strange: 98% of it is the planets' and just 2% the sun's. On the other hand, Struve pointed out another fact that had to have cosmogonic significance. Spectroscopic measurements of rotational velocities showed that the hotter and more massive stars rotate at high velocities (100–500 km/sec), while all stars of lower mass and temperature rotate at low velocities. The distinction, which is quite clear-cut, occurs at spectral type F5. The rotational velocity of the sun at the equator is only 2 km/sec. However, if all the planets of the solar system were to join the sun, by the principle of conservation of angular momentum the sun would accelerate, just as a ballerina or a skater spins faster by bringing her arms close to the body. The sun's rotational velocity would increase 50 times, thus reaching 100 km/sec, which is the velocity of most of the hotter stars. According to Struve, then, the distribution of angular momentum among the planets lowers the rotational velocity of the central star, or, if you like, stars that rotate slowly, namely, those of the later spectral types, announce the existence of planets in their neighborhoods.

At this point the theory that planets form by accretion appears more universal; furthermore, by applying Struve's conclusion, we can estimate the number of existing planetary systems. According to a conservative estimate, in our galaxy there are 40 billion stars of spectral type later than F5. Hence there could be 40 billion planetary systems. If each system consisted of about 10 planets, as should be the case on the average, there would be 400 billion planets in our galaxy alone. The number would be even higher if we assumed that planetary systems also formed by different mechanisms—for example, in the case of double or multiple stars. Finally, one could add the systems created by the tidal effects due to the close encounter of two stars. But as Struve himself calculated, the chance of close encounters is so low that there could not be more than 2 or 3 such planetary systems in the whole galaxy.

COMPLETING THE PANORAMA

THE EARTH-MOON SYSTEM
The planets of the solar system formed about 4.6 billion years ago. The moon was born 100 million years later. Why?

Three entirely different theories have been advanced to explain this fact. The first, formulated at the end of the last century by G. Darwin, son of the famous naturalist, holds that the moon formed from a detached piece of the earth. This theory has proved untenable on many grounds. One of the major difficulties is the different chemical composition of the two bodies, which has become even more evident with the recent lunar explorations—rocks from the earth and the moon differ in the percentages of their various components. In addition, the transfer of a piece of the earth to the lunar orbit is not easy to explain from the point of view of dynamics.

The difference in chemical composition, which has contributed to the rejection of Darwin's hypothesis, is the basis for the second theory—capture. In this view the moon formed like the planets in another part of the solar system and then, chancing to pass near the earth, was captured by our planet, and by its gravitational attraction was compelled to circle around it. This eventuality is possible but extremely unlikely because it places severe constraints especially on the mass and the orbit of the body to be captured. If the mass of the moon had been just a little larger or a little smaller and its orbit had been just a little more or a little less inclined to the earth's orbit, the moon would have collided with the earth or escaped into space.

The third theory is the most comprehensive; it is supported by astronomical observations and by the findings of the recent space explorations; and it is the one that best fits the formation process of the earth. According to this theory, the moon formed from a swarm of (more or less large) bodies in orbit around the earth. As W. M. Kaula and A. W. Harris emphasized in 1973, the formation of the moon from a large swarm of particles orbiting the infant earth is the most acceptable assumption from a dynamical point of view, and the chemical differences between the earth and the moon must be explained in the context of a dynamically correct model. In the past ten years this has been the approach of Kaula and Harris themselves and many other scientists—in particular, E. L. Ruskol, who, on the basis of all assumptions and results, has given the theory a comprehensive and convincing form.

At the time of its formation 4.6 billion years ago, the earth went through a very active phase of accretion. It was during this period that our planet gathered around itself the swarm that was to give rise to the moon. According to Ruskol's calculations the particles accumulated especially on the side of the swarm facing the earth, and many fell on it. The evolution of the swarm occurred rapidly, in a much shorter time than it took the earth to evolve. First minor bodies formed by accretion, by the same mechanism as the planets. At that time the solar system in the making was not as empty as it is today, and in addition to the larger bodies in more or less circular and concentric orbits there were smaller bodies revolving around the sun in elliptical orbits that crossed the others. Many of the small satellites forming around the earth were certainly destroyed by these intruders and transformed into debris that for the most part enriched the swarm.

Some of the bodies growing within the swarm escaped destruction and, having reached a certain size, became indestructible because all the bodies that could still collide with them were too small to shatter them. In practice, only a limited number of satellites with radii over 400 kilometers must have survived. In turn, they combined to form a few protomoons capable of withstanding the bombardment of intruders for more than 100 million years.[29]

It has been calculated that the accretion processes within the circumterrestrial swarm lasted only 1,000 years. Since it took the earth about 100 million years to evolve, almost certainly there was a very long period in which the forming earth was surrounded by various protomoons. According to calculations there were 4 to 6 of them at distances ranging from 19,000 to 383,000 kilometers. Eventually they joined and 100 million years after the formation of the earth the moon was born of their union (figure 37). The determination of this date is based on dynamical considerations and on the age of the moon rocks brought back by the astronauts.

We do not know at what distance from the earth the moon formed.

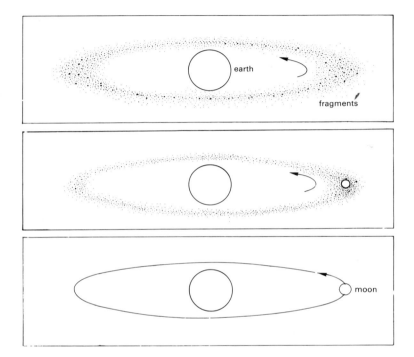

Figure 37   Formation of the moon from a swarm of particles orbiting the earth. These drawings are much simplified, and one must visualize some processes not shown here but mentioned in the text—namely, the earth still in the process of being formed, the formation of some minor protomoons, the destructive action of external bodies, and the formation of the moon (shown alone in the bottom drawing) from protomoons traveling different orbits and coming from different swarms. [From *Le Scienze* (July 1972)]

It was certainly much closer than it is now; hence by Kepler's third law it must have revolved much faster around the earth, which, in turn, was rotating much more rapidly about its axis, so that days and nights alternated at a much faster pace. The lunar month and the earth day gradually lengthened as a result of tidal friction on the ocean floors and solid tides inside the earth. This process is still going on and the day continues to lengthen, as demonstrated by paleontological research and by a comparison between ancient and current astronomical observations.

The early history of the moon is as yet not too well understood. Initially it seems to have had a surface temperature around 700°C. Little by little the temperature increased, mainly because of the energy released by radioactive processes, and 400 million years later it was more than 1 100°C on the surface and even higher in the interior. At that point melting of the lunar material must have been completed. Then, fairly rapidly, the outer layers cooled down and a solid crust formed, which continued to be bombarded by the myriads of smaller bodies that were still roaming the solar system and the earth's neighborhood. Dating of the lunar craters shows that 4 billion years ago, when the moon was 500 million years old, the rate of bombardment was 1,000 times greater than today's much slower pace, which has been going on for 2 billion years. About 3.9 billion years ago, at the end of the most active period, large bodies fell on the moon, forming the basins of the maria (figure 38); they were not filled by lava until much later, 600 to 700 million years later (figure 39), depending on the cases. Since then the lunar surface has not changed very much (figure 40). The fall of meteorites has added a few more craters here and there. A very slow process of erosion is going on due to the solar wind and the micrometeorites, far more numerous than their larger brothers. But apart from this, everything has remained the same. The dark lava fields that you can see with the naked eye from your home window are the same as 3.3 billion years ago, and the lighter regions pitted with craters are still those that formed about 4 billion years ago.

THE SATELLITES OF THE OTHER PLANETS

Beside the earth, other planets have or have had satellites, but their origins and fates have not always been similar to the moon's. When applying the swarm theory to the various planets, a distinction must be

Figure 38   The side of the moon facing the earth as it would have appeared around 3.9 billion years ago right after the fall of the large bodies that formed the basins of the maria. Observe the ring structure of the huge basin that once covered by lava will become Mare Imbrium. In the center of the basin is the large crater that partly submerged by lava will later look like a gulf, the Sinus Iridum of today. [From D. E. Wilhelms and D. E. Davis, *Icarus* 15 (December 1971)]

Figure 39   The same hemisphere of the moon as it would have appeared 3.3 billion years ago. By then lava had invaded the basins forming the maria, and the moon looked very much as it does today (see figure 40). The most significant difference is the absence of some large craters, such as Copernicus, Eratosthenes, and Tycho, which would form later from the occasional fall of large bodies. [From D. E. Wilhelms and D. E. Davis, *Icarus* 15 (December 1971)]

Figure 40   The face of the moon as it appears today. The most interesting features for the sake of a comparison with the previous figures are Tycho (11), Copernicus (17), Eratosthenes (18), Sinus Iridum (21), and Mare Imbrium (22). The reconstruction of the former faces of the moon by Don E. Wilhelms and Don Davis shows that the greatest changes occurred in the first billion years of the moon's life and that its appearance has changed very little since then. [From D. E. Wilhelms and D. E. Davis, *Icarus* 15 (December 1971)]

made between the planets that accreted from dust and the larger ones that formed from gas and dust. The first group includes Mercury, Venus, the earth, and Mars. The first two rotate at very low velocities (Mercury in 58.65 days, and Venus in 243). Evidently, they were subject from the beginning to a strong slowdown due to the tidal effects of the sun, which had a negative effect on the survival of satellites. If around Mercury there formed a swarm that gave rise to satellites, all those with radii over 500 kilometers at distances of less than 97,000 kilometers fell on the planet. The same thing happened to the satellites of Venus with radii over 1,000 kilometers at distances up to 250,000 kilometers. In sum, it is possible that in the past Mercury and Venus also had swarms of particles from which some protomoons developed, but the latter soon fell onto their respective planets, which never succeeded in acquiring a major satellite revolving in a stable orbit.

The two satellites of Mars (figures 41 and 42) could have formed in the same way, but their irregularity and the fact that their craters are very big compared with the size of the satellites themselves indicate that in this case they are fragments of larger bodies. On the other hand, experiments carried out in 1977 by the Viking 1 and Viking 2 orbiters have shed new light on the Martian moons. Phobos, surveyed from a distance of only 80 kilometers, has been found to have a very low density (1.9 grams per cubic centimeter), which suggests that it must contain large amounts of volatile substances like water. R. H. Tolson and his coworkers believe that Phobos did not form either in the region of Mars or at the same time, but came from the asteroid belt. This hypothesis appears to be confirmed by spectroscopic observations, also made by the US probes, which have revealed the satellite's remarkable similarity to the darker asteroids and carbonaceous meteorites. The latter are objects rich in carbon and water and are quite different from the majority of Martian and earth rocks.

As if this were not enough, the other satellite, Deimos, surveyed and photographed from a distance of just 23 kilometers, looks completely different from its companion in that it has a smoother surface erratically strewn with boulders. Undoubtedly, the origin of the two moons of Mars is not yet very clear and reopens the door to the capture theory, at least for this planet.

This possibility is even stronger in the case of the minor satellites of Jupiter and Saturn discovered by the Voyager 1 and Voyager 2 probes

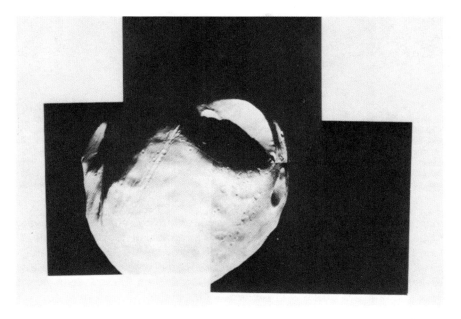

Figure 41   A view of the irregular surface of Phobos, one of the two satellites of Mars, in a mosaic obtained by combining photographs taken in 1977 by the Viking 1 orbiter. The smallest visible features are 20 meters across. [Courtesy NASA]

between 1979 and 1981. Some of them are less than 100 kilometers across and are not even comparable in size to the planets' major satellites, which are of the order of magnitude of earthlike planets. Thus it is possible that at least some of them are captured planetesimals. As far as the major satellites are concerned, in 1980 E. L. Ruskol formulated a swarm theory that takes into account the fact that when they formed there was a large amount of gas in that region in addition to dust.

EVOLUTION OF THE PLANETS AND SATELLITES
The solar system is born! with its planets, its satellites, and its comets ejected far out in space, whence afterward a few will return to their birthplace. From now on, for billions of years, these bodies will continue to revolve around the sun, each following its own evolutionary path. Not all of them will have a varied and brilliant future. The moon, Mercury, some of the satellites of Jupiter and Saturn, and all of the

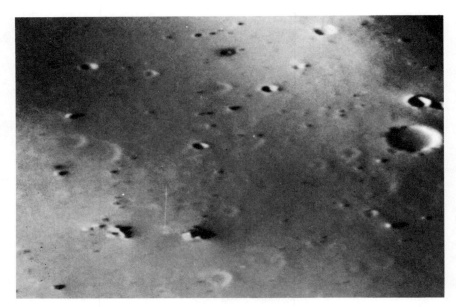

Figure 42   A detailed view of the other Martian satellite, Deimos, also photographed in 1977 by Viking 1. The smallest visible features are 3 meters across. Notice that the surface of Deimos is flat and very different from that of Phobos. [Courtesy NASA]

minor bodies are not massive enough to have retained an atmosphere capable of protecting them and modifying their surfaces. Thus they still show today traces of the final phase of their accretion, preserving the craters made by the bodies that fell last and that no atmospheric agent has ever erased. Apart from some minor changes, due, for example, to lava eruptions, they have not undergone substantial transformations. They lived an intense life for only a brief period right after their formation, then remained almost inert, like fossils.

On the other hand, there are extremely active planets and satellites, such as Saturn, Jupiter, and Jupiter's satellite Io, but theirs is an activity that recalls that of the sun in the sense that everything is constantly changing but basically repeating itself, sometimes according to predictable cycles. Perhaps these bodies, too, change and evolve in the long run, but if they do, we are not aware of it, partly because we do not know them very well, particularly from the evolutionary point of view.

Finally, there are bodies like Venus, the earth, and Mars that above a solid soil have an atmosphere they themselves built and modified and, with the possible exception of Mars, a warm or hot interior. Both the atmosphere and the interior, variably hot, produce continual changes that since the formation of the planet have kept on reshaping it as an artist shapes and reshapes a piece of clay. Among these bodies, the earth is perhaps the most active and certainly the best known, and on the earth we shall now stop.

THE EARTH

The orbits of Mercury, Venus, the earth, and Mars are separated by very large distances, 50–100 times the radii of their spheres of action. This wide separation makes for great stability in their motions around the sun and tells us that their present orbits cannot be very different from those traveled by the original planetesimals.

While the other planets were forming, the earth grew around its embryo by sweeping up and collecting the dust along a ring 76 million kilometers wide whose internal edge was 119.7 million kilometers from the sun and the external edge, 194.5 million kilometers from it. In this manner, in the first 40 million years, or perhaps between 20 and 60 million years, the earth accumulated a substantial mass. Then the accretion process continued more slowly, but in any case in 100 million years the protoearth already had reached 97–98% of its present mass. The rest of the mass was added, ever more slowly, in the millions, billions of years that followed. Thus after 100 million years the process of accretion was just about completed and the terrestrial globe could be said to have formed. It was formed from dust.

All we see today—plains, cities, mountains, animals—comes from that dust, from the particles of metals, silicates, and ice crystals floating in space, small and light like the specks of dust dancing in a beam of light. In those corpuscles was our earth.

But after 100 million years of accretion the earth no longer is scattered dust. Through the joining of countless particles and some intermediate processes it has condensed into a nucleus; it has grown, melted, and amalgamated. In brief, it has merged into a single body; it is a planet about to become a world—a world that after many vicissitudes will become, for a time, our world.

THE FIRST ASPECTS OF THE EARTH

The solar system has formed, and with it a planet of particular concern to us—the one we inhabit. For us, the earth not only is the most important planet in the universe but also is the most interesting because it is the only one whose history we can reconstruct and the only one we know that harbors the extraordinary phenomenon of life.

THE STRUCTURE OF THE EARTH AND THE FORMATION
OF THE CRUST

By studying the propagation of seismic waves we have found that at present the earth can be schematized into four parts (figure 43). The outer part of the planet consists of a thin crust whose thickness is much greater under the continents—20–90 km (kilometers)—than under the oceans (5–12 km). Below the crust there is a region called the mantle, extending to a depth of almost 2,900 km and composed of iron and magnesium silicates. Further down is the core, which is divided into two parts: an outer liquid layer about 2,300 km deep, consisting of nickel, iron, and a little sulfur, and an inner solid part with a radius slightly over 1,200 km and composed solely of nickel and iron.

The temperature at the center of the globe is much higher than on the surface. As we move downward it increases by 3°C (3 degrees centigrade) every 100 m (meters), but this occurs only near the surface; in the mantle the increase is slower. At the base of the mantle the temperature is believed to be 1,800–3,900°C, while at the center of the earth it might reach 4,000°C. This internal temperature seems enormous to us since it is of the same order of magnitude as the temperature of the solar photosphere from which, despite the great distance, we receive so much heat. If we had not measured it by scientific means we would never have suspected it, living as we do on the cold surface of the earth. It is due to this high temperature that at a great depth we find a liquid layer. Nevertheless, the inner core is solid because the tremendous pressure prevailing there is not balanced by a substantial increase in temperature.

This structure dates to the time of the planet's formation. Even though it may not have been born hot, the earth is believed to have warmed up moderately as the aggregating minor bodies fell and converted their gravitational energy into heat. In addition, relatively short-

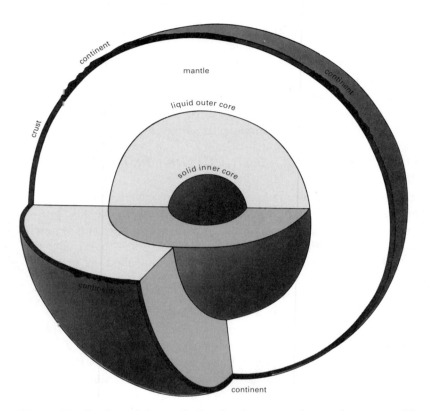

Figure 43   Section of the earth showing its current internal structure. [From M. Ageno, *La comparsa della vita sulla Terra e altrove,* Loescher, 1978]

lived radioactive elements such as aluminum 26 must have been produced by the supernova that exploded in the general area shortly before the formation of the solar system. These elements, certainly present in the material from which the earth originated, heated the forming planet by emitting energy through their rapid decay. Today they no longer exist because they have turned into stable elements, but the internal temperature of the earth is kept high by other, long-lived radioactive elements (uranium, thorium, potassium 40) also present in the original nebula.

Thus, during or shortly after the condensation of the protoplanetary material into a single body, the earth warmed up. Initially it must have been homogeneous in composition in the sense that each of its forming bodies, or at least a certain volume of the primordial material, contained the various elements mixed together. Melting caused the heavier elements, such as iron and nickel, to sink to the bottom, while lighter elements like basalt remained higher up, and still lighter ones like the silicates came to the surface. Naturally all the material was in a molten state. Ascending and descending currents formed in the magma due to differences in temperature from one region to another and in the density of the molten materials. Gases were released in the interior and, seeking a way out, flowed upward through the fluid matter. Thus the surface of the earth must have looked like a vast ocean of bubbling and smoking lava. But no one could have seen this red-glowing globe from space because it was enveloped by a thick, turbid, poisonous atmosphere that by reflecting and scattering the light of the early sun may have given the earth, still in its hellish labor, a light and luminous appearance similar to that of Venus in our sky.

All this must have occurred more than 4.5 billion years ago, when the surface temperature was over 800°C. As the temperature fell to that value or below, which is believed to have happened in a very short time, a solid crust began to form here and there, becoming increasingly more extended and thicker. When the temperature fell below 100°C, water made its appearance, no longer as vapor but in liquid form, as rain. The formation of water perhaps was a fairly early event and no doubt contributed to the acceleration of the cooling process. Nevertheless, the picture did not change very much. There were sheets of water —rivers, or rather floods—that carried rain water, but it was rivers and rains of boiling water that ran over fiery rocks whipped by winds as hot as the air from an oven.

The crust was very thin, and now and then it would break, making way for mighty volcanic eruptions and new outflows of lava. It was often struck and sometimes rent by the very abundant meteorites of that time, last remnants of the material left over from the formation of the planets. Even after it had become more resistant, the crust continued to be struck by meteorites, and, covered with craters large and small, it took on the appearance of the bright lunar regions or certain regions of Mars and Mercury. This landscape was later erased by erosion and perhaps by other phenomena that we shall discuss shortly. By virtue of its internal heat and its atmosphere, the earth, unlike the moon and Mercury, became an increasingly more active, that is, an alive planet that could heal its wounds and in time even erase the scars, a planet that slowly but continuously would be changing its face.

Although we have a pretty good idea of what occurred on the earth, and in the surrounding space, right after its birth as a planet, we have not found a single relic from that distant past. This does not mean that we have not tried. Considering what the earth was like at that time, the only things one could hope to find intact are the original rocks. The oldest rocks known to date, located in western Greenland, are 3.7 billion years of age. However, they do not appear to have been part of the first terrestrial crust because they are sedimentary rocks, formed from the detritus of older rocks that had been removed by erosion and carried away by the first rivers, which most likely deposited them in some sort of primitive sea. Thus the first crust must have formed before 3.7 billion years ago, and this takes us back to that undefined time between 4.5 and 4.0 billion years ago (probably closer to 4.5) that we had already adopted on the basis of the evolution of the protoearth.

THE OCEANS
Thus 4 billion years ago, perhaps earlier, the outer part of our planet had solidified. We know that the internal material was uniform in composition and distribution and that the crust was formed from the lightest components, which had separated during melting and come to the surface everywhere. Hence the entire planet should have been covered by a fairly homogeneous and even crust. In other words, there should have been neither mountains nor large depressions such as the current ocean basins. But in fact the crust does not have the same level everywhere; it is thicker under the continents, and thinner under the oceans. Furthermore, the two crusts are different in composition; the continents are made essentially of granite, calcareous rocks, and other light material, whereas the ocean floors consist essentially of basalt. Finally, and most surprisingly, their ages are also different. The continental rocks may be thousands of millions of years old, while the rocks of the ocean floors are much younger, a few million years of age, at most 160 millions.

These facts are very strange. But in 1977 H. Frey proposed an explanation that fits well with the history of the earth's neighbors: Mercury, Mars, and the moon.

As we have seen, around 4 billion years ago the moon was struck by large bodies whose impact caused the formation of circular basins. Mercury and Mars also were scarred by the intense bombardment of bodies the size of asteroids, the largest remains of the material from which the planets and satellites had formed. If they struck the moon, it stands to reason that such bodies also fell on the earth, which was in its immediate vicinity and, moreover, much larger and more massive. Extrapolating from the lunar basins, that is, from the hits received by the moon, Frey calculated the hits received by the earth and concluded that over 50% of the solid crust that had spread uniformly over our planet was shattered by the heavy bombardment. The basins thus formed were inundated by basaltic lava and later became the places where water collected, while the surviving crust constituted the land masses. If this theory is correct, the modern division of the surface was determined early in the earth's history. According to Frey's calculations, the current ratio (70% oceans to 30% dry lands) also was determined by those impacts. Thus many contradictions and differences appear to be resolved. Of course, the impact craters do not coincide with the present oceans. The basalt that spilled into the chasms made by the impacts came out of the mantle 4 billion years ago, whereas the basalt of the ocean floors is no older than 160 million years. Evidently something has changed since then; indeed, it is still changing. According to Frey, even this process started with those initial impacts. More about this later.

Regardless of how they formed, the large basins have since had an inseparable companion—the ocean. It cannot have taken long for water to fill the cavities formed in the crust since sedimentary layers dating to 3.5 billion years ago have been found in various parts of the globe. This proves that the ocean already had existed for some time. Water existed even earlier. Perhaps it coexisted with the molten magma since, as we can see today, an equilibrium is established between the water dissolved in volcanic magma and the water vapor that forms above it. Hence not even the hot molten rocks could totally destroy the stores of water incorporated into the planet at the time of its birth. Furthermore, studies show that the ocean is very stable chemically and physically, and its main characteristics cannot have changed very much over time. In this respect the primitive ocean must have been similar to ours, especially after the complete cooling of the earth's crust.

But it was very different in one respect. Something was missing—life. Try to imagine it for a moment. Think of the modern seas and try to see them without a fish, without seaweed, without a branch of coral, without the myriads of microorganisms that populate each of their drops. That is how the underwater world was then—deaf, blind, mute like the deserts of Mars or the moon. But this very environment was about to become the cradle of the life that one day would fill the entire planet.

THE ATMOSPHERE

While the oceans do not seem to have changed very much since their formation, except in salt content, the earth's atmosphere underwent more than one radical change, starting from mixtures of unbreathable gases and ending with the present one that until a few years ago was the best suited to our ecological balance.

The first atmosphere may have been a mixture of the gases found in the nebula from which the sun and the solar system had just formed. It should have been rich in hydrogen, helium, and other noble gases (neon, argon, krypton, xenon). But perhaps it never quite formed because it was dissipated into space by the solar wind during the accretion process of the earth. The fact that even the heavier gases like krypton and xenon were blown away makes us think that this atmosphere never did envelop the finished planet. Their percentage in our atmosphere is only a ten-millionth of their abundance in the sun, which must correspond to their abundance in the primordial nebula. The fact that they are present in such reduced proportions means that they were blown away before the forming earth had reached its present mass; they were too heavy to escape later from the earth's gravity and could not remain as compounds since they do not react with other elements.

The second atmosphere formed from the outgassing of molten rocks, and long after the formation of the solid crust it continued to be fed by the numerous volcanoes in full activity. There is some disagreement on its composition. The presence of water vapor (H_2O), nitrogen (N_2), and carbon dioxide (CO_2) is universally accepted. More controversial is the presence of ammonia (NH_3), carbon oxide (CO), and methane (CH_4). Hydrogen certainly was released, but it cannot have been an abundant constituent, and its presence in the atmosphere must have been transitory because, being light, it would have rapidly escaped into space. Finally, everyone agrees about the lack of free oxygen.

In the course of hundreds of millions of years this atmosphere evolved into the present one. M. H. Hart has attempted to reconstruct the various phases of its evolution by programming a computer with all the necessary data, including interactions with soil and oceans, so as to obtain the variation in the gases every 2.5 million years. Among the numerous initial atmospheres tested by the computer, Hart chose the one that would evolve into the present one, with its current constituents in their actual percentages. According to this reconstruction we find that initially the main constituent of the earth's atmosphere was carbon dioxide, which subsequently diminished rapidly by reacting with siliceous rocks, while most of the water vapor condensed, forming the oceans. Nitrogen was scarce, free oxygen all but nonexistent (according to all the experts), and consequently from about 4.3 to 2 billion years ago the principal components of the earth's atmosphere were methane and other carbon compounds. They disappeared 2 billion years ago when nitrogen already had become prevalent and a fundamental element— oxygen—had long since appeared, though in minute amounts.

We shall see later how the atmosphere has further evolved in the past 2 billion years. Now it is worth noting that from this model of the atmosphere Hart calculated how the average surface temperature and cloud cover varied in the various epochs. In the first 2 billion years of the planet's life the average surface temperature was fairly high (45–22°C versus 7°C at present) and the sky was always overcast. Then with the rapid decline in the abundance of methane, in less than half a billion years the cloud cover fell below the current value. The skies opened up, the sun illuminated and directly heated the soil, and by night, instead of a dark overcast sky, the starry vault appeared. But as yet there were no living beings that could warm themselves in the sunlight; nor were there eyes that could see the stars.

LIFE

It is curious how much the birth of a celestial body like the earth resembles that of a human being. The fetus originates and develops in the mother's womb, nourished and produced by something that, like itself, comes from an external environment. The father has initiated the process of formation by launching one of his elements toward another one that the mother had kept ready for a while. From that moment on her womb represents its vital environment, and furnishes it with the right heat, the food to nourish it, the equivalent of breathing; but it also can transmit disease to it, cause malformations, and even death. After a gestation during which its life had little influence on its person, formed but still enclosed and bound, it will come out with the characteristics transmitted by its parents and developed in the mother's womb. From that moment it will become an independent individual that will govern itself ever more and ever better from its inner self, influencing, with its own will, the new external environment. The latter will still act on it, but the value of the external-internal ratio (apart from energy, provided by food) will be reversed.

Something similar happened with the earth. Having developed from a body that was passing through a cloud of dust, it was formed the way the cosmic environment in which it was developing wanted it: the primeval material determined its composition; the decay of the radioactive elements emitted by a nearby supernova forged its interior; the fall of asteroids carved out its ocean basins. Many, many other factors, such as its dimensions, its distance from the sun, the presence of a large body like the moon in its vicinity, and meteor showers, shaped the earth in a certain way, different from that of its brother planets. At this point the external influences diminish markedly (apart from the supply of energy from the sun), and the earth, shaped by all it has received from its protoplanetary environment, begins to live a life of its own, independent of that environment, not so much because it has emerged from it, but because that environment has changed, indeed almost vanished. Nearly 4 billion years have gone by since then, and now we shall see how the earth has lived through them.

THE ORIGIN OF LIFE

Today everywhere we look we see life: people and animals passing by in the streets; trees; meadows covered with a great variety of grasses and flowers and populated by countless insects; other animals underground; birds and more insects in the air. Even where our sight cannot reach, the air, lands, oceans, and deserts too are filled with life. We do not even need to look around us. Each of us is composed of billions of living particles that together constitute an entity richer than its individual components—our organism.

This living world, which to us seems so normal and obvious, has not existed from the time of the earth's formation. We do not know whether there are similar worlds elsewhere on other planets; we do not even know exactly what it consists of. Nevertheless, if we wish our voyage in time to give us the most comprehensive view possible, these questions must be addressed. Not only is life, even if it exists only on our planet, an integral part of the universe, but also and above all it represents the ultimate development of matter, its transformation into something that we feel no longer seems, and perhaps no longer is, matter. We shall begin from the last point; first of all, that is, we shall try to define in the most general way what is meant by life.

In the first place it should be made clear that life is not a *state* but a *process* that unfolds according to some fundamental, material characteristics that distinguish it sharply from the world of inanimate matter. Also, in defining what is meant by a living being we shall make no reference whatever to "spiritual" attributes; in other words, our definition must apply to all known life forms, from an amoeba to a human. Finally, it is obvious that such a definition cannot be considered either absolute or definitive, particularly because it is based on the only life we know, the terrestrial.

Our experience tells us that a living being may be defined as an organism that

1. takes matter from the environment and decomposes it, using the products to manufacture the macromolecules of which it is made,
2. produces reactions that release chemical energy, which is partly used and partly stored,
3. grows, and
4. reproduces itself.

Naturally a living being is such only if it satisfies all four conditions. A star, for example, can grow by collecting the interstellar medium that it encounters in space, but it is not a living organism because it does not satisfy the other conditions.

The most elementary living organism we know is the cell.[30] Do not be misled by the term "elementary," which suggests something extremely simple and thoroughly understood. No less than five volumes have been devoted to the cell, so far, in the popular science series in which this book first appeared in Italy. But it is the most elementary of all living beings, and there is no doubt that the first inhabitants of the earth must have been unicellular organisms.

Like a human, the cell also satisfies the four conditions stated above. Cells are divided into two broad categories: eucaryotes and procaryotes (figure 44); these terms, derived from the Greek, denote, respectively, cells with a clearly visible nucleus enveloped by a membrane and cells without a distinct nucleus. The nucleus contains the hereditary apparatus that enables the cell to reproduce itself.

A cell is composed of proteins, nucleic acids, polysaccharides, and lipids. Proteins are long chains of amino acids found in every part of the cell and are of great importance to the manufacture of structure and chemically active substances essential to life. Polysaccharides are molecular chains formed from the molecules of glucose and its derivatives; best known are starch and cellulose, which differ only in the way the glucose molecules are strung together. Glucose, which generates the polysaccharides, is produced in chlorophyll-containing plants from carbon dioxide, water, and sunlight. Oxygen is released in this reaction. Starch and sometimes cellulose are the essential nutrients for all herbivorous animals. Polysaccharides and lipids serve to supply energy to the cell, besides being an integral part of some of its organs. For example, some lipids, along with proteins, form the outer membrane of eucaryotic cells.

More complex and extraordinary is the role of the nucleic acids, which are of two types: deoxyribonucleic and ribonucleic acids, commonly known as DNA and RNA. We shall consider only the first, DNA, a macromolecule shaped like a double ribbon (figure 45) whose strips are coiled helically around each other. Its great importance lies in the fact that stored in it, as on a computer tape, is all the information relating to the cell. Just as we can make an exact copy of the tape that will have the computer perform the same operations, the cell can duplicate its DNA and this duplicate can generate and regulate a new cell as the original did (figure 46).

Briefly, this is how the DNA regulates the cell. Every chemical reaction occurring in the cell needs a special protein, called an enzyme, which is characterized by a particular sequence of amino acids. Such a protein does not take part in the reaction; yet the reaction cannot take place without it. In practice it represents what chemists call a "catalyst." The synthesis of each of these enzymes is regulated by a piece of the DNA molecule known as a gene.

Proteins, nucleic acids, polysaccharides, and lipids—the basic con-

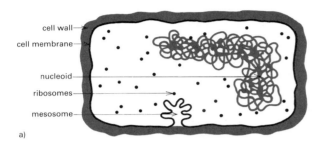

a)

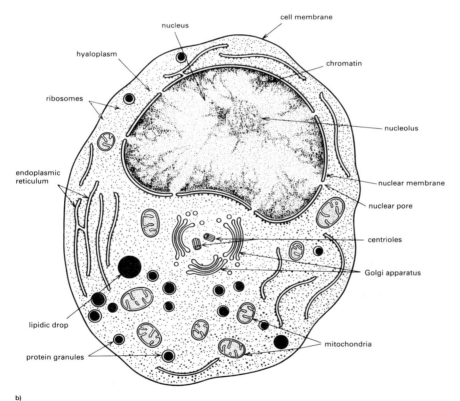

b)

Figure 44   Diagram of a cell, the simplest living organism: (a) the primitive procaryotic cell; (b) the more advanced eucaryotic cell, with a well-defined nucleus. Structures not mentioned in the text are shown here to underline the complexity of the latter type of cell. [(a) from E. O. Wilson et al., *La vita sulla Terra*, Zanichelli, 1977; (b) from M. Bessis and J. P. Thiény, *Intern. Rev. Cytol.* 12 (1961):199]

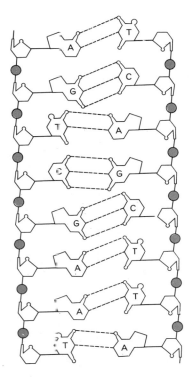

Figure 45   Diagram of a section of a DNA molecule showing parallel polynu-
cleotide chains. Each chain (for instance, the left one) consists of alternating
molecules of phosphate (filled circles) and deoxyribose sugar (pentagonal
shapes); a molecule of a substance called "base" is attached to each sugar
molecule. The bases are four: two purines, adenine (A) and guanine (G), and
two pyrimidines, thymine (T) and cytosine (C). The sequence of bases in the
chains determines the genetic information carried by the DNA molecule. While
the order may be random in one chain (and in fact varies from organism to
organism), it is restricted in the other chain in that the pairing of the two (to
form a complete DNA molecule) can occur only between given bases—for
instance, A and T or G and C.

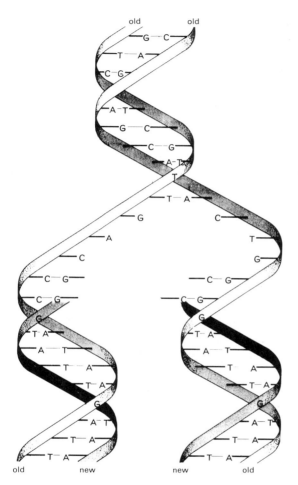

Figure 46   Diagram of a DNA molecule showing the parallel chains twisted into a double helix and the process of duplication.

stituents of the cell whose main functions we have just seen—originate from a number of molecules called "prebiotic" because they are the material used to manufacture those components of living beings, like the DNA molecule, that already seem to be alive.

What are these prebiotic molecules? When and how did they first form? And how did they give rise to the first cells?

The time has come to return to the earth that we left newly formed and devoid of life. As far back as the 1920s the Russian biochemist A. I. Oparin maintained that given certain conditions and sufficient time, life, that is, organized matter, develops from nonliving matter as a result of an evolutionary process that is mostly chemical in nature. According to Oparin, such conditions existed in the atmosphere and seas of the early earth, where the first organic substances, such as hydrocarbons, cyanides, and their immediate derivatives, growing more complex and combining into larger, chain-shaped molecules, formed a mixture of the most disparate organic substances. Dissolved in the primitive seas, they created the so-called primordial soup from which life originated.

This idea, advanced in the same period by the English geneticist J. Haldane, gained widespread acceptance over the years. Then in 1953 a chemistry student, S. Miller, decided to test its validity with a laboratory experiment. Using two glass containers connected by pipes, Miller simulated the conditions of the primitive sea and atmosphere by forcing a mixture of hydrogen, water vapor, ammonia, and methane to circulate through a liquid water solution. Had there been nothing else, the "terrestrial atmosphere" would have certainly reached a thermodynamic equilibrium with the "ocean" and nothing would have happened. But in the environment of the early earth there was (as there still is) a continuous input of energy from the sun's ultraviolet radiation, lightning discharges, radioactive substances, and volcanic eruptions. Accordingly, Miller had electric sparks continually produced in the glass containing his "atmosphere." After a week he noticed that the liquid simulating the ocean had become cloudy. Subsequent analysis revealed to his amazement and satisfaction that the liquid contained several amino acids, the building blocks that living organisms use to manufacture proteins. Miller's experiment has since been repeated with many variations in the composition of the atmosphere and the type of energy source. All such experiments have consistently led to the synthesis of numerous prebiotic molecules.

In sum, Miller-type experiments performed in the 1950s and 1960s have demonstrated that the simple precursors (like amino acids) of biological macromolecules (like nucleic acids and proteins) could have formed in the primitive terrestrial atmosphere. This had to mean that an ocean fairly similar to the present one and an atmosphere composed of water vapor, hydrogen, ammonia, and methane, and with the input of energy that certainly was not lacking, could produce fundamental prebiotic molecules like amino acids, fatty acids, carbohydrates, purines, and pyrimidines. From these came the basic components of the cell (proteins, DNA, etc.), but exactly how the proteins and DNA essential to the process of cell duplication and regulation were first formed is not yet clearly understood. Yet we know that prebiotic molecules could have formed on earth; life does exist, and the transition, though still obscure, must have occurred at some point. Let us search the distant past for the earliest traces of life.

THE OLDEST EVIDENCE

Until recently the earliest living organisms were believed to be those found by the Nagys in the sedimentary rocks of Swaziland in southeast Africa. They were simple, roundish structures a few thousandths of a millimeter in diameter, perhaps unicellular blue-green algae that lived about 3.3 billion years ago.

Today's paleobiologists, more cautious in their approach, categorize the earliest signs of life in three ways according to their degree of reliability, namely, "possible," "probable," and "evident." According to this definition the oldest "possible" traces of life are those found in Greenland and Zimbabwe in 1979, going back 3.7 billion years. The oldest "probable" traces were found at Pilbara, Australia, in 1978 and are 3.5 billion years of age. The oldest "evident" traces are findings in Canada, announced in 1975, and in Belingwe, first studied in 1975, which date to 2.7 billion years ago. Almost all of these traces of life are found in particular sedimentary formations, previously underwater, known as stromatolites, which still form today from layers of blue-green algae.

Blue-green algae and anaerobic bacteria (capable of living in environments devoid of free oxygen) appear to be the earliest forms of life on earth. They are both unicellular organisms consisting of a procaryotic cell that reproduces by dividing into two identical cells. Blue-green algae (as well as anaerobic bacteria) still exist today, and though called blue-green they come in all colors. They are capable of living even in the most unfavorable environments, and perhaps this explains why this form of life, one of the first to appear on earth, has survived until today, about 3.5 billion years.

The oldest known fossil about which no doubt has ever been raised was found in the Transvaal in a rock 2.2 billion years of age. It is a filamentous alga, clearly an evolved form of life. It may have taken evolution more than a billion years to produce it, and it is not much of a feat considering that in those many years it only accomplished the transition from isolated single-celled organisms to assemblages that did not constitute a multicellular organism but rather a colony of similar cells, independent and strung out side by side.

Perhaps in all that time life did not even progress from procaryotic to eucaryotic cells. As reiterated recently,[31] the existence of eucaryotes older than 1.4 billion years has not yet been proved. The great change appears only 1.2 billion years ago. The oldest cells of which we know that are certainly eucaryotic date to that time and are found in the Beck Spring Formation in California.

The formation of eucaryotes (cells endowed with a true nucleus), which to the best of our knowledge was accomplished in about 2.3 billion years from the time life first appeared, was the greatest innovation in the history of our planet. The advent of this type of cell marked the beginning of multicellular organisms and sexual reproduction; some evidence of the latter was found in the first large assemblage of procaryotes discovered in the Bitter Spring Formation in Australia, dating to 900 million years ago. Both events were very important for various reasons.

All multicellular organisms are composed of eucaryotic cells and have two great advantages over single-celled organisms. They can form groups of specialized cells that control different functions, and therefore have a much greater range of possibilities than single-celled organisms. And they can grow larger than other organisms and hence feed on them. The maxim that "big fish eat little fish" dates from that time—

in a broad sense, of course, since there were no fishes as yet. It is still valid today, though for different reasons.

As far as sexual reproduction is concerned, it is also found in single-celled organisms, but it is normal and common in the multicellular ones. It has some marked advantages over asexual reproduction and the most important is the following. As I mentioned earlier, the fundamental step in reproduction is the duplication of the DNA molecule (or molecules), which carries the genetic information. This duplication does not always result in a perfect replica of the information contained in the DNA of the original cell because errors occur in gene replication, either by chance or as caused by chemical agents or radiation. More generally, in the genes of an organism there can occur changes, called mutations, that normally are passed on to all the descendants. As a result, little by little, over many generations, mutations change the characteristics of the descendants. In the case of sexual reproduction the genetic heritage of the parents is mixed at random. Hence the offspring will not be an exact copy of either parent and the process of change of the species will be greatly accelerated.

This process is the basis of evolution, and life on earth depends on it. Consider a group of organisms that reproduces sexually and has characteristics specifically suited to its environment. Should the environmental conditions suddenly become unfavorable, these organisms might die and the species would become extinct. But if the change is gradual, their descendants may have time to diversify to the point that a part of them will be able to adapt to the new conditions and survive. In other words, sexual reproduction causes numerous, continuous changes in the progeny: the individuals that by chance are ill suited to the new environment die off, while those that are better suited survive and find themselves in equilibrium with the new environment, something that would not have happened if there had been no mutation. This is the process known as *natural selection*.

It is this process that ensures the survival of the species and promotes its evolution as the natural environment changes—the organisms that survive are either those that can live in the most diverse environments, like the blue-green algae, or those whose chance mutations enable them to adapt to the new conditions. Naturally mutations occur also in organisms that reproduce asexually, but sexual reproduction accelerates genetic variation, thereby providing a much faster response to the evolutionary processes due to environmental changes.

At this point we perceive the necessity of death in all its crude reality. As we shall see better later on, the earth is constantly changing, and so is the environment of every species. Hence evolution is a necessity. The individual, even if eternally young, could not change beyond some very narrow limits. By the mechanism of life and death the species can change and adapt to new environments, and even give rise to new species more advanced than the old. Thus the purpose of death is not the same as that of life: we do not die to make room for others like ourselves. The reason for the passing of the generations is altogether different. The reason is evolution—the individual must die for the species to evolve.

If the great advantage of sexual reproduction concerns the species, what is in it for the individual? And what makes the individual act as nature intended? We all know the answer—love, a force of nature based on biochemical processes that drives animal and vegetable species regardless of physical pleasure, which may be absent. This force often produces the most exhilarating, the most revolutionary, the most constructive moment in the life of a living being. It is through love that after meeting and attracting each other, two beings of the opposite sex can reach a mutual devotion capable of annulling their individuality by culminating in a single aspiration, in which both unconsciously sense the generations to come through the feeling of eternity, of time standing still, that only love can give. In human beings love has been excited and transformed to the point of becoming an essential part of the subconscious, causing joy and pain, life and death, and changing into attachment to one's children, love for all beings and things, mysticism, and even hatred. It has become an essential part of our lives, in literature, art, religion, and morals. But we are not the only ones capable of love. Millions of years before it was sung by the poets of the "dolce stil novo," love was experienced by creatures different from us and from our close ancestors. Although to us they may seem monstrous and repulsive, those creatures, too, felt such an attraction for the opposite sex that, like us, they were willing to sacrifice themselves for the sake of mutual possession, through which would be born a progeny from which, after all, we also descended.

But it is time to return to the fossils of a billion years ago, among which we have not yet found any animal.

THE FIRST ANIMALS

In 1947, while conducting a geological survey of sandstone formations at Ediacara, Australia, R. C. Sprigg discovered a large number of unusual imprints, unmistakably fossils (figure 47). The paleontologist M. F. Glaessner of the University of Adelaide became interested in the strange fossils and in a series of expeditions in that area, beginning in 1957, found a great variety of them. Subsequent studies revealed that they were imprints left on the mud of the ocean floor by soft animals (devoid of shell or skeleton) that had completely disintegrated after death. Unfortunately, the rocks at Ediacara could not be dated. But meanwhile similar fossils had been found elsewhere in fragments of rocks that were dated at 700–600 million years ago. Thus at that time there were organisms in the seas that by moving or lying on the bottom left their imprints, soon covered by sediments that protected those tenuous traces of life. Later on, as the waters withdrew, the sediments hardened, thereby preserving the imprints until we could interpret them.

They were organisms 10,000–100,000 times larger than those that had lived also in Australia 2.8 billion years before and no longer consisted of a single procaryotic cell but of millions of eucaryotic cells. Above all they were animals, the first animals to appear on earth. Almost certainly they were primitive crustacea, jellyfishes, and worms. The first marine organisms with a hard covering appeared in large numbers and in a great variety of forms around 570 million years ago. Prominent among them were the trilobites, of which 3,900 different species are known. They dominated the seas for almost 200 million years, and for a while were the most advanced animals on the planet; then slowly they became extinct. The graptolites were another large group of organisms living in that period; they diminished greatly in number after 100 million years and were believed to have become extinct more than 300 million years ago until a type of graptolite, perhaps representative of their last descendants, was fished out of the sea in 1948. These new creatures were not colonies of unicellular vegetables like the blue-green algae that had formed stromatolites, but complex organisms with a digestive tract, prehensile organs, sexual organs, a nervous system, means of locomotion, and eyes and other sensory organs that enabled them to "personalize" their lives instead of just vegetating as all previous organisms had done.

Around 450 million years ago—250 million years after the first animals had appeared—the sea already contained most of the principal components of contemporary marine fauna. There would still be significant extinctions and some partial substitutions: the brachiopods would decline and be almost entirely replaced by mollusks; the trilobites would be replaced by crabs and similar animals; shelled cephalopods would be replaced by shell-less octopuses and squid. But by that time life already had made the great leap from the primitive animals that had left a few imprints at Ediacara. This "modernization" of marine

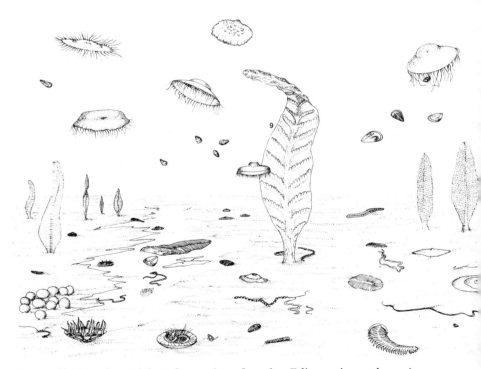

Figure 47   Imprints (right) of organisms found at Ediacara in southern Australia. The drawing (left) is an attempt to reconstruct the marine environment where this fauna developed. [From M. F. Glaessner, *Scienzae Tecnica,* 1974]

fauna, begun 150 million years after the first animals had appeared, took about 100 million years. The large and varied fauna that populated the seas of that epoch included fairly advanced echinoderms, such as the familiar starfishes and sea urchins. Relatives of theirs were evolving into vertebrates, from which humans would eventually descend.

THE ATMOSPHERE AND THE CLIMATE
As I mentioned earlier, the atmosphere of the primitive earth did not contain free oxygen. Its appearance and increased production were a consequence of life and at the same time permitted a greater development and more rapid evolution of life forms. By the well-known pro-

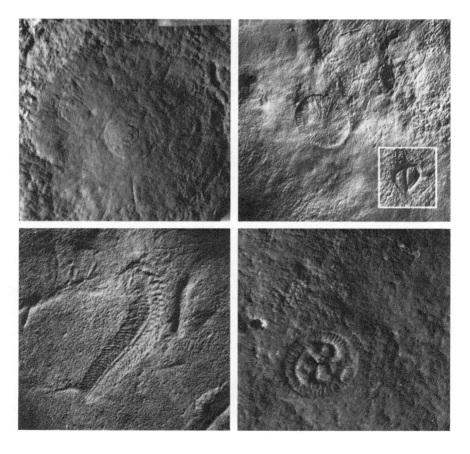

cess called photosynthesis, green plants use the energy of sunlight as absorbed by chlorophyll to build carbohydrates from water and carbon dioxide, releasing the oxygen in water as free oxygen.

This process, essential to the production of food and energy in vegetables, already was taking place in blue-green algae about 3 billion years ago. At that time carbon dioxide was plentiful, while oxygen may have been present only in trace amounts. Because of the lack of free oxygen in the atmosphere, the ozone layer at high altitude had not yet formed, and lethal ultraviolet radiation from the sun, which today is absorbed by the ozone, could penetrate to the surface of the earth. This made it impossible for life to develop on land. But it could exist in the water, starting from those depths that the sun's ultraviolet rays, blocked by the overlying layers, could not reach. Blue-green algae, which were very resistant to ultraviolet radiation, were able to live near the surface, producing a considerable amount of oxygen that was released into the air.

The first turning point occurred when the oxygen content of the atmosphere reached 1% of its current value. Under those conditions some microorganisms were able to turn from fermentation to respiration by absorption of oxygen. With that percentage of atmospheric oxygen, moreover, the sun's ultraviolet rays began to penetrate seawater only to a depth of some 10 meters. Hence organisms vulnerable to ultraviolet radiation now could live and develop below that level, that is, in relatively shallow waters. Meanwhile oxygen continued to be produced, and when its abundance in the atmosphere reached 10% of the current value, there began to form the ozone layer at high altitude that henceforth would shield the soil from the sun's ultraviolet radiation. From that moment on living beings would be able to develop and proliferate also near the ocean surface and on dry land.

Thus in approximately 3.5 billion years the earth's atmosphere became more rarefied and transparent, protective and breathable, practically the atmosphere we have today. The event had two momentous consequences in the history of life: an enormous increase in the number and variety of marine life forms, which occurred about 600 million years ago, and the emergence of life on land between 440 and 400 million years ago (figure 48).

The climate is closely connected with the atmosphere. And since climatic variations have far-reaching effects on the life of the planet, both locally and globally, it is important to learn if and how the climate

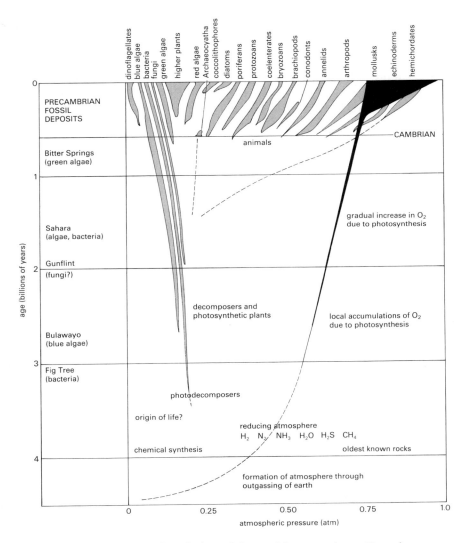

Figure 48   Formation and evolution of the earth's atmosphere. Note that oxygen began to form with the contribution of the first photosynthetic plants and has been increasing since, particularly in the past 400 million years. [From E. O. Wilson et al., *La Vita Sulla Terra,* Zanichelli, 1977]

changed in the past. Unfortunately it is a very difficult problem. The climate depends, first of all, on the latitude and elevation of a place, but also on the density, composition, and transparency of the atmosphere; on the shape and distribution of the continents, which affect warm and cold ocean currents; on mountain ranges; and on astronomical factors, such as variations in the inclination of the ecliptic (the plane of the earth's orbit around the sun), in the eccentricity of the orbit, in the position of the perihelion (the point in the earth's path nearest the sun), and in the earth's rotational velocity, which was higher in the past. It is evident that many of these factors, like variations in some of the earth's orbital elements, will affect the climate, not just locally, but all over the world. The trouble is that the variables are many, and not all of them can be calculated. And even when they can be calculated, it is impossible to work back to the remote ages that are of greatest interest to us. In the case of the earth's orbital elements, for example, even with the help of modern computers we have not been able to go further back than a few million years. Thus attempting to reconstruct from current data the climate of past ages is a difficult and risky business. The most profitable approach, as in our pursuit of early life, is to investigate the geologic record. This type of exploration has reached far into the past, revealing striking and unsuspected phenomena.

COLD WAVES
Various studies, particularly of the flora and fauna of the past, indicate that, as a rule, the earth's climate has always been mild and rather warm. At times, however, there have been cold spells that radically change the face of the planet. Besides the harshness of the climate, and as a consequence of it, some conspicuous phenomena occur during these periods. Large amounts of water turn into ice, forming immense glaciers that spread over the poles and all those regions that are cold because of latitude and elevation. With so much water withdrawn from the ocean to form the glaciers, the general level of the seas falls, while the weight of the ice pushes the land downward, partly sinking it into the warm, soft mantle underneath. As a result new lands emerge all over the planet while others are submerged by the sea, subsiding under the weight of enormous ice sheets that transform the landscape, covering and erasing every form of life.

The last time this phenomenon occurred, the glacier that formed over Scandinavia spread as far as northern Germany. On Lüneburg Heath south of Hamburg one can still see, resting gently on the smooth terrain, scattered boulders that the ice had carried there from thousands of kilometers away and that were left behind when the glacier retreated. In the same period, which ended only 10,000 years ago, the region of Canada where today rises the city of Toronto was under a layer of ice 1,500 m thick. The sheet of ice covering Hudson Bay was even thicker, 2,400 m. The enormous weight of this icecap pushed the continental crust into the earth's mantle for some hundreds of meters. When the ice melted the North American crust slowly began to rise, and its upward movement is recorded in a series of terraces sloping to the current beach and corresponding to the shorelines of various periods. The highest terrace is 150 m above sea level and carbon dating shows that it formed 7,000 years ago. This means that as the ice melted, the coast rose at the average rate of 2 cm (centimeters) per year.

These extraordinary cold spells have been called glacial epochs (or eras). The way to discover them is very simple—all one needs to do is to find datable evidence of glaciers where they should not have occurred, for example, in low-lying regions at great distances from the geographic poles. The most common evidence is provided by characteristic marks left by the ice on preexisting rocks and by the so-called tillites, a type of sedimentary rocks very similar to those that are produced even today by glaciers and icebergs. In practice, the work of the geologist is not so simple. It involves digging all over the world, uncovering the evidence, dating it, and other complicated work we shall not go into.

At least five glacial epochs have been recognized, of which only the last one, not yet ended, is well understood. The first datable glacial era seems to have lasted from 2,300 to 1,700 million years ago—a span of time so vast that it may have comprised four glacial epochs. Another well-documented one occurred between 700 and 600 million years ago and lasted 50–100 million years. A third, lasting about 10 million years, began 450 million years ago. The fourth one was much longer; it started approximately 340 million years ago and lasted about 100 million years. The fifth and last one began 2 million years ago and is still going on. Naturally it is the best known, having left abundant evidence, not only in rocks and fossils, but also in the habitats of our predecessors who evolved during that period.

One of the most interesting aspects of a glacial era is that it comes in waves, called glaciations. In the current era, for example, cold spells have alternated with periods of mild or warm weather. One such period occurred between 130,000 and 115,000 years ago. After the last glaciation, about 15,000 years ago, the temperature began to rise and we are currently in an interglacial period of mild climate, although the temperature has started to fall again.

These continuous fluctuations of the temperature during a glacial era, discovered also for the past ones, though in less detail, must be taken into account when attempting to explain its causes. On this subject, more theories and hypotheses have been proposed than I can possibly mention. In a book published in 1961, M. Schwarzbach listed as many as 54 hypotheses advanced up to 1950. Many had been already discounted, but until a few years ago various possibilities still were being discussed, including variations in the amount of radiant energy emitted by the sun, passage of the solar system through clouds of cosmic dust as it revolves about the galactic center, sudden though inexplicable variations in the rotational velocity of the earth, and a gigantic buildup of the antarctic icecap.

Reality is perhaps simpler and less catastrophic. As early as 1941 M. Milankovič had suggested that we were dealing with a phenomenon somewhat similar to the seasonal cycle, but longer and more complex. It is well known that the seasons are caused by the annual variation in the inclination of the sun's rays and in the amount of time the sun is above the horizon, both of which determine the amount of solar radiation absorbed daily by the surface of the earth. This variation results from the inclination of the earth's axis, which remains constant throughout the year, so that the exposure of the earth to solar radiation changes during its annual revolution about the sun and different places experience the phenomenon in turn. If the inclination of the earth's axis, which we have assumed to remain constant for one year, were to change slowly over the years, a very slow seasonal cycle would be superposed on the annual cycle. The same thing would happen if there were changes in two other geometrical conditions that affect the insolation of a place, that is, eccentricity of the earth's orbit and inclination of the orbital plane. Depending on whether all these "seasons" add up negatively or positively, there would be exceptionally warm or exceptionally cold periods.

Milankovič's results did not gain widespread acceptance, but in 1976 his theory was reproposed by numerous scientists with the support of geologic evidence. Stratigraphic samples taken from the floor of the Indian Ocean had made it possible to reconstruct the changes in the earth's climate for the past 450,000 years, and this research showed that the worldwide oscillation of the climate not only had been periodic, but also exhibited the three periods calculated by Milankovič on the basis of orbital variations.

Since variations in the earth's orbital elements always take place, one still had to explain why glacial eras are exceptional events of a few million years (with widely fluctuating temperatures) in a normally mild climate. The answer seems to be that under normal conditions the oscillations of Milankovič's "seasons" do not have a marked effect; but if for some reason the average temperature of the planet falls, the oscillations are amplified, causing warm spells to alternate with cold spells —in other words, the cold waves characteristic of a glacial era. This mechanism has been reconstructed by V. Ya. Sergin through computer simulations. The causes for the lowering of the world's average temperature may differ from one glacial era to another. They are known for the current one and will be discussed when our journey in time takes us close to the present.

Now we should return to our planet, where, since we left it, evolution is occurring at an accelerating pace and in an ever more complex way.

LIFE EVOLVES

THE GEOLOGIC ERAS
In our exploration of the earth's past from its birth to the appearance of the first animals we have measured time in years, but only a few decades ago we could not have been so precise. Up till then our knowledge of geologic time was very rough because radiocarbon dating had not yet been used; moreover, the age of the earth was not known with any certainty. As late as the 1950s one still read in textbooks that the period we have just explored lasted 1,200–2,800 million years, whereas in reality it lasted 4,000 million years. The last 600 million years of the earth's history were better known because a great many events, particularly life, had left abundant records in the rock strata that permitted a fairly accurate relative dating. Even so, as we go back to the begin-

ning of geology and paleontology, early in the eighteenth century, we find that time estimates become more and more imprecise. One thing seemed clear from the start, however: Where there was no sign of inversions, one could assume that the rock strata had accumulated one on top of the other with all they contained and that each layer was older than the one above it. Thus it was possible to assign an epoch to each layer and to determine its order among the other epochs and to know from the fossil remains of vegetables and animals contained in the various strata which had lived before and which afterward. This is why geologists divided the history of the earth into eras and periods with names that are sometimes evocative because they recall the dominant event (Carboniferous, Cretaceous, etc.), but that more often are strange and difficult to remember, having been derived from the location of the finds (Cambrian, Silurian, etc.).

Eras and periods defined by their boundaries expressed in years do not make much sense. What is important is their placement in the scale of time, their relative duration, and above all their fundamental characteristics. Let us take an example from history—the Middle Ages. Everyone has a concept of that period, perhaps somewhat personal but certainly well defined, that is independent of the exact day it began or ended, which in fact are least typical of the entire period. This is why geologists prefer to speak of eras and periods, and from now on we shall have to do the same, although I shall continue to give dates in years whenever possible. In any case the relation between the two systems is furnished by table 4, which one should always keep in mind.

The era that goes from 600 to 225 million years ago is called the Paleozoic. Its first period was called the Cambrian after Cambria, the medieval name of Wales, where the rocks of this period were first studied. The geologic era preceding the Cambrian, that is, the era we have just explored, is called Precambrian Time and comprises 87% of the earth's history.

THE EVOLUTION IN THE WATERS

By the end of the Ordovician period, about 450 million years ago, marine fauna had undergone a tremendous development. It already included forms that would last to this day; among them, two types of animals, arthropods and vertebrates, would vie for supremacy, first in the water and then, much later, on land and in the air. The arthropods

Table 4   Geologic eras

Beginning date (millions of years ago)	Era	Period
0 —	Quaternary	Olocene
2 —		Pleistocene
7 —	Cenozoic	Pilocene
25 —	or	Miocene
40 —	Tertiary	Oligocene
55 —		Eocene
65 —		Paleocene
135 —	Mesozoic	Cretaceous
195 —		Jurassic
225 —		Triassic
280 —	Paleozoic	Permian
350 —		Carboniferous
400 —		Devonian
440 —		Silurian
500 —		Ordovician
600 —		Cambrian
3,300 —	Precambrian	
3,700 —		
4,600 —	Formation of the earth	

were already present with all their subgroups, from the very abundant trilobites, which would later become extinct, to crustaceans, Chelicerata, etc. The vertebrates were represented by the first fishlike animals, the Agnatha, or jawless fish. They were a strange sort of fish, with no fins, a cartilaginous rather than a bony skeleton, and an armor covering them from head to mid-body and, in some cases, almost to the tail. Most likely, tens of millions of years earlier they had had armorless progenitors that had been easy prey to the trilobites, mollusks, and other marine animals. To defend themselves from predators they developed an armor and migrated, first to brackish lagoons, then to fresh waters, moving inland up the ancient rivers and settling in lakes and marshes.

Meanwhile the arthropods had produced some unpleasant animals, the eurypterids, that, improperly but effectively, are often called water scorpions because of their appearance and habits. They may have existed since the Cambrian in a variety of types and sizes, but one type found in abundance in the Silurian, between 440 and 400 million years ago, was up to 2 m long, had a pair of toothed claws, four pairs of "walking legs," and a pair of flattened appendages that served as oars. Stalking their prey, these predators over millions of years followed the route of the armored fish, moving from the sea to lagoons, then to river mouths, rivers, and inland waters. The fish had lost their haven. Nevertheless, the fossil record shows that the inland waters of the Silurian already harbored an abundance of fish of a great many varieties, not counting those that had returned to the sea. Possibly there were several waves of migration to and from the sea.

The fish of this period still have an armored head but begin to show two fins, one on the back and the other under the tail. They had three eyes and one nostril, all on the same side, and on the basis of morphological imprints it is speculated that they may have had electric organs, which they used to stun or kill their attackers, after the fashion of modern electric eels. All these data suggest that they led a lazy life, lying on the bottom, camouflaged by sedimentary material, with only their eyes and nostrils showing through. If attacked, they could stun or kill their enemies with a powerful electric shock. Being no bigger than 30 cm, they fed on small prey, sucking them as lampreys do. Of course these bottom dwellers were not the only fish of the time. In the waters above them swam various types of fish with an eye on each side of the head. Some were armored; others had no armor but a skin bristling with spines; still others, with neither armor nor spines, were endowed with large fins and entrusted their safety to their quickness in evading predators.

Up to 400 million years ago fresh waters were the great nursery where fish underwent their long evolution. At this point begins the next period, the Devonian, which lasted 50 million years. At the beginning of this period the agnathans were still increasing, and to them were added other armored fish. Meanwhile, since the end of the Silurian, the first jawed fish, the placoderms (now extinct), had begun to branch out from the primitive agnathans. Though armored, they were quite agile because their armor was well articulated and connected to the body. They had movable jaws with cutting plates that made them extremely dangerous. As if these attributes were not enough, they grew to enormous sizes. The largest and most fearful of them, Dinichthys, reached 9 m. These formidable fish had developed like the others in fresh waters and then migrated to the sea. There they destroyed not only a great number of their predecessors but also many arthropods, including the pseudoscorpions that had dominated the seas up to that time. Thus, toward the end of the Devonian, about 360 million years ago, the arthropods were defeated by the vertebrates, which became the masters of their common environment. Fate had reversed their roles. And progress did not stop here. In the Devonian seas we find numerous members of a new class, the Chondrichthyes (fish with cartilaginous skeleton), whose descendants include sharks, skates, and rays.

In those same seas, initially abandoned by the fish, the rest of the fauna had undergone various changes. The graptolites had died out; the lovely starfish had appeared, and though in decline, the trilobites were still prominent along with the cephalopods, which had spawned the first ammonites.

Among the strange animals populating the seas and inland waters we still have not met any of them we commonly call fish today: an aquatic animal, spindle-shaped, with fins and scales, and above all with a bony skeleton, the very spine that gives us so much trouble when we eat fish and normally ends, along with the head, at the edge of our plate. This type of fish constitutes the class of the Osteichthyes, or bony fish, which, as the name implies, have a bony structure. They appeared around 390 million years ago, began to develop in the course of the Devonian, and little by little filled the seas and the fresh waters. The bony fish were so successful that today they are twenty times more numerous than all other kinds and comprise more species than all the other vertebrates put together.

Thus the development of marine vertebrates was marked by three major stages: the agnathans, primitive fish that did little more than vegetate; the placoderms, more mobile but still weighed down by armor; and the bony fish, freed from armor and endowed with greater agility and initiative. The last ones would overwhelm the others, setting off on a road that would take some of their descendants to a point no creature of that time could have foreseen—not even we ourselves, if transported there with our intelligence but without our current knowledge.

LIFE EMERGES FROM THE WATER

To the modern mind, life is so tied to the land that we never give much thought to the creatures of the sea except when we enjoy them for dinner or amuse ourselves by peering at them through a diver's mask. Yet the underwater world has always been filled with life. Life started there, and there it flourished for billions of years, evolving into the rich fauna we have just met. During those long years the land was lifeless—a desert of rocks, debris, and sand first shaded by a perennial cloud cover and then flooded by lethal ultraviolet radiation from a sun shining in a generally cloudless sky. Meanwhile, however, the oxygen released by marine organisms was enriching the atmosphere and gradually making it breathable, and eventually enough ozone formed from the oxygen to shield the soil from the sun's ultraviolet rays. At that point the land became habitable and life emerged from the water to invade the continents. First a few pioneers came ashore, as in all invasions; then there was a very slow expansion, still tied to the water, and finally, in an accelerating evolutionary process, life freed itself of its watery habitat and spread into most of the terrestrial environments.

Conditions favorable to the emergence of life on land occurred in the Silurian period, between 440 and 400 million years ago; that is when the first pioneers landed. Unlike the landing of Christopher Columbus in America as portrayed in paintings, this momentous event was not attended by any emotional, self-conscious ceremony. As the fossil record shows, the first colonizers were scorpions, relatives of the eurypterids, pushed by some inexplicable necessity. Herbivorous animals such as the myriopods, ancestors of today's millipedes, probably went ashore at the same time; otherwise the scorpions would not have had anything to eat. On the other hand, vegetable life was also spreading inland during this period. The process started toward the middle of the Silurian with a veritable explosion of green marine algae. Then, still in the Silurian, strange vegetables halfway between algae and terrestrial plants appeared at the edge of the water and especially in swamps. Their stems consisted essentially of side-by-side, intertwined tubes and had an outer layer of waterproof cells that protected the plants from evaporation. They had no roots. The base of such a plant sat in the water, absorbing it and transporting it to the aerial part. From these primitive plants came the Psilophytales, precursors of ferns and all higher plants.

To understand how these first plants worked and the evolutionary step they represent, let us stop here a moment and look at a modern plant. A watery solution containing minerals is drawn from the soil by the roots and then distributed to the cells of the leaves by a system of woody vessels called xylem. This liquid is called raw sap. After reacting with the products of photosynthesis, which occurs in the leaves, the elaborate sap flows downward to nourish the plant through a system of living cells, the phloem, located in the outer part of the plant. This differentiation of conductive elements is already found in the fossil Psilophytales that lived about 400 million years ago; though primitive, the vascular system is there. These plants lacked roots and had a rhizome, lying horizontally in the mud or wet earth, from which rose smooth, leafless shoots that forked at the top; each tip carried a sporangium, the sack of spores used for reproduction. Some of these plants, larger at the base, reached a height of 3 m and might have looked like real trees.

This great advance, and it was only a beginning, was due to a new cell, the tracheid, which perhaps had developed several million years before. The vascular tissue made from these cells has a rigid structure capable of holding a plant upright and is the main element for carrying water with dissolved minerals tens of meters from the soil. With this tissue large plants could develop without being supported by water or being in direct contact with it. These new plants could live far from the sea and swampy areas. The stage was set for the development of a large and varied vegetation all over the continents.

The Psilophytales are considered a class of the tracheophytes, the type that gave rise to all high-stemmed plants living or extinct. In the second half of the Devonian, between 380 and 350 million years ago, the tracheophytes developed roots that by capillary action absorbed water from the subsoil even where there was none on the surface. They also developed more or less complicated branches and leaves of various types. Such plants resemble today's ferns and horsetails but were much larger. By the end of the Devonian we find plants, the pteridosperms, that reproduced by means of seeds rather than spores, and primitive gymnosperms, the cordaites, that reached a height of 15–18 m.

All these plants formed the first forests, where spiders and scorpions lived and multiplied together with the primitive insects a few millimeters long that lived on rotten vegetable matter. The development of underbrush and organic soil completed the transformation of the once barren and lifeless land, providing an inviting ecosystem for a great many species of animals that would soon be invading the continents.

Let us return then to the animals. The arthropods, which appeared in the seas long before the vertebrates and were later defeated by them, had their revenge by being the first to walk on the land. In reality there was no competition but rather a slow evolutionary process. It is only today that we see a competition, and we cannot help cheering for "our team," that is, the vertebrates, the remote ancestors not only of humankind but also of familiar and well-loved creatures like dogs, cats, and horses. If the vertebrates had never left the water we would not be here today, and neither would those that in our inveterate anthropocentrism we call animals—from lions to apes, from birds to deer. Around 400 million years ago, while the arthropods were settling on the land, the vertebrates were still in the water. What were they waiting for? Perhaps just the push of necessity.

Among the bony fish living early in the Devonian, some 390 million years ago, there were some strange types that still have a few living descendants and belong to the order of the Dipnoi, or lungfishes. Their main difference consisted in having developed a pulmonary system in addition to gills (Dipnoi means "two breaths") and four paddlelike fins that enabled them to waddle about. When a drought dried up the swamps in which they lived, they burrowed into the mud, breathing air through their "lungs" until the water returned. If the water was late in returning, they died. We have been able to reconstruct these small tragedies partly because we observe them in their descendants, and partly because chunks of hardened mud containing their fossil remains have been found in Devonian deposits. Despite such intriguing attributes as primitive lungs and fleshy fins, the dipnoans never made the transition from water to land.

Even more interesting are the Crossopterygii, an order of fish that was thought to have died out 70 million years ago. It was quite a surprise therefore when in 1939 a member of a living species was fished out of the sea. This order included the family of the Rhipidistia. Beside lungs, these fish had teeth and paired, limblike fins with bones similar to those of frog legs. When the waters in which they lived dried up, they crawled over the land in search of other ponds and marshes. Natural selection favored those with sturdier limbs and more efficient lungs, and at the end of the Devonian, about 350 million years ago, the rhipidistians disappear and we find the first amphibians. Not all of the crossopterygians followed this route. Many returned to the sea, including the remote ancestors of the fish caught in 1939, and their primitive lungs turned into an air bladder, which made it easier to swim at different levels.

Meanwhile the conquest of the land on the part of the fish was marked by new successes. In the late Devonian appear the Ichthyostega, primitive amphibians with large skulls having sensory organs for orientation (typical of fish) but also endowed with legs and lungs. They reached a length of about a meter.

There was still a long way to go in the colonization of the land, which would require many adaptations, and even today we are not wholly emancipated from the water, at least at the start of our lives. Think of the frogs, for example, which lay their eggs and begin their lives in the water, or the liquid in shelled eggs, or the amniotic fluid in which we spend the first nine months of our lives. But by the end of the Devonian, about 350 million years ago, the vertebrates had taken the first great step on dry land. In the great evolutionary race they had caught up with the arthropods.

CONTINENTAL DRIFT

We have just seen that between 400 and 350 million years ago life emerged from the water and invaded the continents. The question is,

Were they the same continents we know today, with the same mountains, the same plains, the same rivers and lakes? And were the continents in the same geographic positions, separated by oceans and seas that were the same as today's? We already know that it cannot be so because a great many fossils of marine animals—indeed, layers of fossils—have been found in continental deposits, often in areas like the Alps and the Andes that today are mountainous. In part, this is explained by the fact that the formation and melting of icecaps during the glacial eras caused the continents to sink and rise again, so that some regions were alternately submerged by the sea and left dry. But this is only part of the story.

Since the seventeenth and eighteenth centuries, when exploration produced the first accurate maps of the continents, people have noticed that the east coast of South America and the west coast of Africa are so complementary that if one pulled the two continents together they would interlock like pieces of a jigsaw puzzle. This fact caught the attention of a young German scientist, Alfred Wegener, while he was glancing through an atlas that had been given him as a gift on Christmas day 1910. Convinced that this remarkable "fit" could not be just a coincidence, Wegener conceived the idea that those two continents, and perhaps all the others, once were joined and then had drifted apart, moving to their current positions. After searching for geologic, astronomical, and paleontological evidence of such a movement, he concluded that the continents had moved and were still moving, and that some 200 million years ago they were all united in a single land mass, a supercontinent he named Pangaea. Wegener published the results of his theory in 1912, in 1915 (in its definitive form), and in 1929 (the updated version). In November 1930 he died on a glacier in Greenland, where he had gone to measure the thickness of the ice, sacrificing his life to save his companions.

Wegener's cause was enthusiastically embraced by a few scientists but was opposed by the majority until, in the 1960s, his ideas were modified and confirmed by new findings. Thus today we can have a unified view of the earth that by means of a single phenomenon explains such disparate and strange problems as the presence of the same fossil faunas and floras in widely separated places, the mystery of Atlantis, the formation of oceans and mountains, the cause of earthquakes, and the origin of volcanoes.

CONTINENTS IN MOTION

Between 1968 and 1971 the opponents of the theory of continental drift received the last crushing blow. In January 1968 a geologist from New Zealand took to the United States a fossil fragment of a bone found on a mountain in Antarctica about 700 km from the South Pole. The fossil was identified by E. H. Colbert as a piece of the jaw of a labyrinthodont, a primitive amphibian that lived in mild or tropical climates between 340 and 200 million years ago. The rock containing the fossil was dated at about 200 million years of age. At that time, therefore, the climate of that region, so very cold today, had to be tropical or at least mild. The fossil find created a sensation and Colbert himself joined a new expedition to Antarctica. On 4 December 1969, in the region of Coalsack Bluff, he found a key fossil, a fragment of the skull of *Lystrosaurus*. This reptile, squatty and stocky, up to 1.5 m long, and herbivorous, lived in streams, rivers and lakes, almost certainly in packs, around 220 million years ago. Earlier on, fossil remains of *Lystrosaurus* had been found in abundance especially in South Africa and India. How could an animal so obviously incapable of negotiating the sea cross thousands of miles of ocean to live on a continent that, assuming it was the same as today, should have been absolutely inhospitable? Colbert's memorable find was only the first in a growing collection. Soon after, more *Lystrosaurus* bones were discovered, and the following year the fossil hunters uncovered the remains of small primitive reptiles and amphibians that previously had been found in the South African fauna of the same epoch. Long before these finds, moreover, fossil specimens of *Glossopteris*, a seed fern that lived between 280 and 220 million years ago in Australia, India, Madagascar, Africa, and South America, had been unearthed in Antarctica too.

The close relation between plant and animal species of different lands was one of the arguments advanced by Wegener in support of continental drift, but the opponents of his theory argued that seeds and spores could be carried long distances by wind and birds and that some animals could manage sea crossings on driftwood. Some even theorized that at one time the continents were connected by land bridges, namely, other continents or huge islands that had later sunk beneath the waves. Last to subside, in fairly recent times, would have been the legendary Atlantis.

The discovery that over 200 million years ago a tropical fauna and

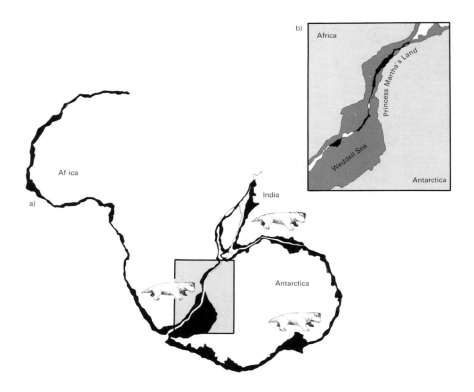

flora flourished on a continent like Antarctica, so very cold today, can only be explained by moving Antarctica next to Africa, Australia, South America, India, and Madagascar. But there are many other arguments that have been brought to bear (especially in the 1960s) to demolish the theory that the continents are stationary and to prove that they drift.

The paleontological argument, greatly strengthened by the discovery of African and Indian fossil animals in Antarctica, had already been brought forth by Wegener for the faunas of other continents. A second argument, advanced long before Wegener, was the matching of the continental contours. Here, too, we have made a big step forward. It can be shown that by appropriately shifting and rotating the continents and islands, their coastlines match quite well. The fit becomes nearly perfect if instead of considering the shorelines, which have changed in so many millions of years due to erosion by the sea, atmospheric agents, and so forth, one looks at the continental shelves underwater.

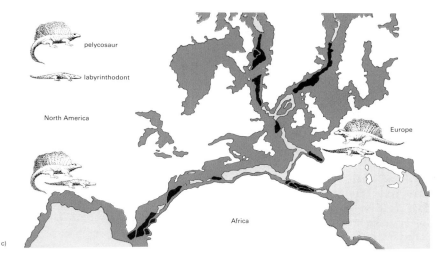

Figure 49   (a) A part of Pangaea reconstructed by fitting together Africa, Antarctica, India, Madagascar, and Ceylon. Observe that the contours of the continents at a depth of 2,000 meters match better than their present-day coastlines. Hence *Lystrosaurus*, whose fossil remains have been found in lands that today are widely separated, at that time was living on a single continent. (b) Enlargement of the region defined by the rectangle—the areas in black are overlaps in the contours; the areas in white are gaps. (c) The fit of the continents in the northern part of Pangaea. Again, the continental shelves at a depth of 1,000 meters (dark gray) match better than the current coastlines; areas shaded black are overlaps; areas of light gray are seas and gaps. Labyrinthodonts and pelycosaurs, animals of which we shall speak later, are found in Texas and Central Europe, today separated by the sea. [From E. H. Colbert, *Animali e continenti alla deriva*, Mondadori, 1978]

Their striking fit is evident in figures 49a and 49b, showing Africa, Antarctica, India, Ceylon, and Madagascar with their present continental shelves to a depth of 2,000 m. It is also evident in figure 49c, which shows the grouping of Africa, North America, Greenland, and Europe. Such a perfect match cannot be due to chance.

The third argument is even more interesting because it has led to unexpected findings. It is based on the study of paleomagnetism, or the earth's magnetism in time past. As you know, the earth behaves as though its interior contained a huge magnet whose axis is nearly coincident with the axis of rotation. The points at which the axis of rotation intersects the surface of the earth are the geographic poles. The magnetic

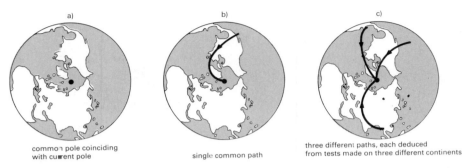

Figure 50   Paths of the North magnetic pole according to the three hypotheses: (a) if there is neither continental drift nor shifting of the magnetic pole; (b) if the pole has shifted but the continents have not moved; (c) if both the pole and the continents have moved. The data have been derived from an examination of the variations in the magnetism of rocks. In (c) the three different paths are due to the combination of the single motion of the pole and the three different movements of the continents—this is the real case. [From D. York, *Il pianeta Terra*, Boringhieri, 1975]

axis deviates from the axis of rotation by only 6°, and it is thought that this inclination has always remained constant or nearly constant. If we travel to the various continents with a compass, no matter where we are the needle always turns to the same point—the magnetic North Pole. Suppose now that we journey back in time with our compass and make the same measurements from the same continents, say every 25 million years. There are three possibilities (see figure 50). If both the pole and the continents remain stationary, we shall always find the pole in the same position from any point on the earth and in every epoch. If the pole moves with the passage of time, the compass needle will point in different directions in turn, showing the movement of the pole over the earth's surface. If all the continents are stationary and only the magnetic pole shifts, the path traveled by the pole must be the same no matter from which continent we measure it. But if the continents also move, the path of the pole will be different depending on which continent we measure it from because the motion of the pole is combined with that of the specific continent.

   Since we cannot go back in time with our compass, this crucial experiment seems impossible. But by a stroke of luck the past has left its own compasses all over the world. All we have to do is find them, date them, and read them. To detect a magnetic field we need the field itself and a magnetized substance that will reveal it by orienting itself along the field's lines of force, just as the compass needle does with the earth's magnetic field.

Extrusive and volcanic rocks expelled in a molten state are not magnetized, but they become so when they cool down to a certain temperature (called the Curie point), which depends on the material. At this point the rock becomes magnetized in the direction of the earth's magnetic field. This means that when an ancient lava flow solidified, it recorded the direction that the earth's magnetic field had at that time. In addition, the solidified rock retains that orientation forever. Hence all we have to do is look for volcanic rocks on different continents, date them, and find their magnetism. The study of magnetized rocks has revealed that the magnetic pole has shifted in time and that its paths over the globe are as numerous as the continents from which measurements are made. These results, corresponding to figure 50c, prove that the continents do indeed move.

Besides providing evidence of continental drift, work on paleomagnetism has given us a better knowledge of the earth's magnetic field. To begin with, we have learned that it exists and has retained more or less the same intensity for at least 2 billion years. Also, that it fluctuates continually and, above all, that it can reverse its polarity, with the magnetic North Pole becoming the South Pole and vice versa. In the past 130 million years the earth's magnetic field has reversed its polarity as many as 171 times, at average intervals of 300,000–500,000 years. We do not know what causes this phenomenon or why the earth should have a magnetic field when not all of the celestial bodies have one. Be that as it may, we can be happy with the results we have obtained thus far since they provide definite proof for a splendid comprehensive theory concerning the structure of the earth's surface and its changes in time.

THE OCEAN FLOORS
Our knowledge of the ocean floors is quite recent. Most of it dates to the post–World War II period when improvements in sounding techniques brought about by the war effort made it possible to obtain accurate profiles of the seafloors. The need to fill a great gap in our knowledge of geography and the wish to solve some outstanding scien-

tific controversies provided the impetus for an extensive exploration of the seabeds, with the result that today we know them almost as well as the continents on which we live (see figure 51).

Contrary to a long-standing belief, the ocean basins are neither flats nor valleys sloping down to their maximum depth at the center. In the middle of the Atlantic, for example, there is a mountain chain that runs more or less parallel to the American and African coasts. This chain, known today as the Mid-Atlantic Ridge, was discovered in 1872, but since it was known only very roughly it remained a mystery, particularly because the oceans were believed to be deepest at the center. As everyone knows, mysteries cannot last very long without turning into myths. Thus the inexplicable presence of a relatively shallow area in the middle of the Atlantic gave new life to the legend of Atlantis, the submerged continent mentioned by Plato.

Much has been learned since then. Similar ridges have been discovered in the other oceans as well; furthermore, they are connected. Indeed, one can say that there is a single ridge that starts from the Arctic Ocean, crosses Iceland, runs through the Atlantic, turns north below the tip of Africa into the Indian Ocean, and then after passing south of Australia, crosses the Pacific and turns north again. Extending for a total length of 64,000 km, it is the longest mountain chain in the world.

But this is not the most extraordinary aspect of the midocean ridges. Whereas the continental ranges culminate in a series of graduated peaks, the sea mounts are topped by a valley, or more precisely, by a steep cleft about 50 km wide and 600–2,000 m deep that runs the entire length of the ridge. Before explaining the significance of these fissures, let us finish our topological survey.

As we have seen, the deepest part of the oceans is not in the central areas, which in fact are occupied by mountain ranges. On the contrary, underwater exploration has revealed that great gulfs, called trenches, open up in the ocean floors near the continental coastlines and the great arcs of volcanic islands. They are long, narrow, and 7,000–11,000 m deep. The ocean ridges and trenches are the sites of the most intense earthquake activity on earth, but with a basic difference. Charting of the epicenters has shown that the earth tremors occurring at depths up to 100 km correspond to the ridges, while the quakes with hypocenters between 100 and 700 km correspond to the trenches (figure 52).

One last startling discovery concerns the thickness of the sediments that blanket the ocean floor. In the Atlantic, for example, there are almost no sediments in the vicinity of the midocean ridge, but moving away from the ridge they become thicker and thicker symmetrically on both sides, reaching hundreds of meters at the farthest points close to the coastlines.

The synthesis of all these findings into a comprehensive view of crustal movement was accomplished by an American geologist, H. Hess. The first draft of his theory was circulated in 1960 in a paper that the author presented to his colleagues as "an essay in geopoetry." And the vision he gave of things was truly one of the most beautiful pages of poetry of nature. While in Wegener's view the continents plowed through the rocks of the ocean floors, Hess envisioned a process of expansion of the ocean floors whereby the continents, being lighter, ride passively on the moving seabed like objects on a conveyor belt.

The mechanism is the following. In certain regions of the mantle, relatively narrow and long, hot material rises by convection, much like hot air rises to the ceiling of a room (see figure 53). In the case of the earth, its outermost layer, the lithosphere, is the ceiling that prevents the hot material from escaping and forces it to move sideways in two opposite directions, pushed by the underlying material that continues to rise. As a result of these opposing motions the lithosphere is pulled apart until it splits, as shown in figure 53. The hot material in the underlying layer, called the asthenosphere, is kept almost solid by high pressure. But where the lithosphere splits there is a marked drop in pressure, the hot material becomes fluid, pours out of the fissure as fiery magma, and forms the first ridge on the ocean floor. The solidified strip continues to be pulled in two opposite directions and eventually splits in two, though not necessarily down the middle. New magma erupts from the fissure, solidifies, and forms another ridge. In turn this splits in two and in time a series of parallel strips forms on either side of the central fissure. Carried by the convective motions of the high asthenosphere, the two sides of the ocean floor spread apart in a slow but continuous motion.

Now the analogy of the conveyor belt is clear: the lithosphere is the moving belt, and the high asthenosphere with its upward and downward convective motions is the roll or system of rolls that moves the belt. The top of the lithosphere is the ocean floor. The continents rest on the lithosphere and are dragged by it, bending it slightly in the same

Figure 51   Map in relief of the ocean floors. It was constructed by Marie
Tharp from bathymetric studies of the Lamont-Doherty Geologic Observatory.
[Courtesy Marie Tharp, South Nyack, NJ]

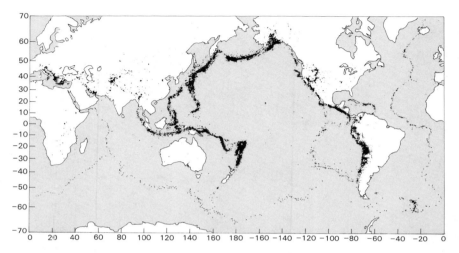

Figure 52   Epicenters of the earthquakes that occurred in this century. The hypocenters range from immediately below the soil to a depth of about 700 kilometers. The quakes corresponding to the ocean ridges originated at depths of up to 100 kilometers, and the others at greater depths. [Courtesy Seismological Society of America]

way that an object will sink into a flexible conveyor belt. Thus the crust, that is, the lithosphere, is generated along the midocean ridges. This process goes on all the time, and since the surface of the earth is not expanding, it means that the crust is being destroyed elsewhere. This is precisely what happens in the oceanic trenches. Here a section of the crust bends and slips diagonally under another piece of the crust, sinking back into the asthenosphere where it is destroyed by the heat, which is the more intense, the farther down the material sinks. As the old ocean floor descends beneath continents or island arcs it forms a downsloping area of seismic activity, and the deeper the earthquakes, the farther away they are from the region where the descent began, which is where a trench opens up.

This explains why there are relatively superficial earthquakes in connection with midocean ridges, caused by the lava that comes to the surface, forming new ocean floor, and deeper tremors where the crust sinks back into the mantle. It also explains why under the oceans the crust is thinner and of a different composition, and why the ocean

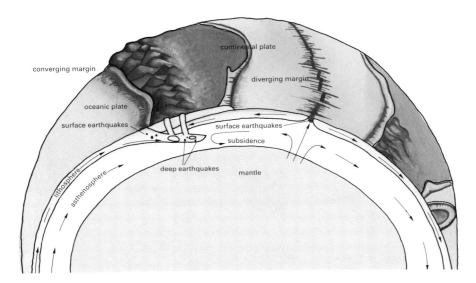

Figure 53   Movements of crustal plates: on the right, the Mid-Atlantic ridge between two spreading plates; on the left, the trench of the subduction zone above which rise the Andes. [From *Science Digest* 1 (1981)]

floors are continually renewed while on the continents there are large expanses of rocks that may be billions of years old.

Hess's brilliant theory of "seafloor spreading" was confirmed by studies of paleomagnetism. Extensive magnetic surveys of the ocean floors carried out in the 1950s had revealed a striking pattern of parallel magnetic bands. In 1963 F. Vine and D. H. Matthews proposed an explanation for the magnetic strips that combined the concept of seafloor spreading with reversals in the earth's magnetic field.

As the lava erupting along a midocean ridge cools, it becomes imprinted with the polarity of the current magnetic field. In time this strip of ocean floor is split in two by a new lava eruption and the two sides are pushed apart. If meanwhile the magnetic field has reversed its polarity, the newly forming strip is magnetized in a direction opposite to the first. This new strip in turn is split in two and pushed aside. After a number of reversals, there will be a series of magnetic bands parallel to the central rise and alternating in their polarity. The magnetic pattern will be symmetrical on either side of the ridge (see figure 54).

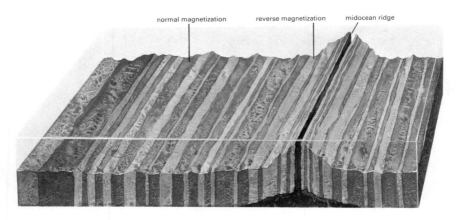

normal magnetization   reverse magnetization   midocean ridge

Figure 54   Magnetic patterns on a section of ocean floor. The strips, parallel to the midocean ridge and alternating symmetrically in their magnetic polarity, record successive reversals in the earth's magnetic field. [From *Il grande libro del mare*, Mondadori, 1978; illustration by C. Di Ciancio, Milan]

Recalling that molten rocks are magnetized according to the field of the time in which they cool, the magnetism of time past, as imprinted in sea rocks, confirms that new ocean floor is continually generated by mantle material that erupts from the rift in the midocean ridge and then slides away, moving slowly but relentlessly in opposite directions. By dating the strips closest to the coastlines we can learn when the original magma emerged from the ridge. In the case of the Atlantic Ocean the strips turn out to be 150 million years old. That is when the Atlantic was born.

PLATE TECTONICS AND MOUNTAIN BUILDING
As we have just seen, it is not the continents that drift over the ocean floors, as Wegener had proposed, but whole blocks of lithosphere that move. In essence, this is the difference between the old and the new concepts of continental drift. Such blocks of lithosphere, called plates, may consist of ocean floor alone or ocean floor and continents. Each plate is defined by its contours, which normally are ridges or trenches. At present there are 6 main plates and about as many minor ones; they are illustrated in figure 55.

The figure also shows in which areas the plates are moving away from each other, with lava eruptions, and in which areas they collide,

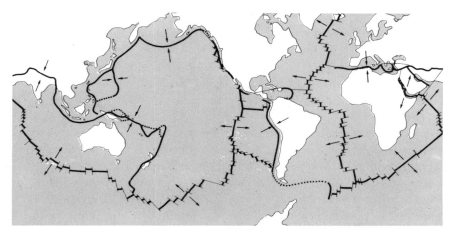

Figure 55   The major tectonic plates; the arrows indicate their movements.
[From *Astronomy* (August 1978)]

ending one under the other. The clash of two plates causes a wrinkling
of the crust, that is, a mountain chain. As long as they are driven by
convective currents in the mantle the plates keep on advancing and
there is no letup in the pressure. This is what is happening in the Pacific with the Nazca Plate, made of seafloor alone; for the past 15 million years, as it descends under the western part of South America,
this plate has been pushing up the Andes. Sometimes we find mountain ranges like the Himalayas or the Urals that rise inside a continent.
This means that they formed from the clash of two continental plates.
Take the Himalayas, for example. Some 180 million years ago peninsular India still was part of Pangaea and lay between Antarctica and
Africa (see figure 49) about 7,000 km from the Asian continental mass.
Dragged by its plate, whose other end presumably was dipping into a
trench along the southern border of Asia, India collided with the continent about 45 million years ago. The penetration of the Indian plate
under the Asian plate caused the Himalayan chain to rise in all its majesty. Here, too, the pressure persists, pushing the Himalayas even
higher.

The sloping descent of one plate under another is often marked by
other phenomena. As the descending plate penetrates deeper into the
mantle, it reaches levels that are increasingly hotter and it begins to

melt. Since the lava tends to rise, sometimes it forces its way upward and breaks through the crust, forming chains of volcanoes that mark the area where deep in the mantle a plate is melting. Volcanoes are thus the most impressive and frightening evidence of the earth's continuous activity, showing up in the critical border zones between plates where terrestrial dynamism erupts and transforms the face of the planet.

Everyone associates volcanoes with death and destruction—everyone but the people who live there. It is true that there is no more desolate sight than a black lava flow just solidified, as barren and lifeless as a lunar landscape. But new life is flourishing all around. Shrubs sprout where only a few decades ago there were rivers of fire. Young verdant pine forests clothe the slopes that were overrun by the magma spewed out by the mountain a few centuries ago. On a volcano, as nowhere else in the world, one has the feeling that life is unvanquishable and that death is only the beginning of new life, a necessary partner rather than a defeated enemy. Seeing a mountain covered with greenery is normal, but seeing lush vegetation rise from the ashes seems like a miracle. It is almost like watching the dawn of the earth, when the fires had gone out and the first living things appeared.

And there is more to a volcano than the life taking hold on its slopes, in its valleys, and atop its craters. The volcano itself is a living thing— restless, unpredictable, tremendous. The volcano is not a common mountain. Mountains also evolve; they are born, grow, and die worn down by erosion. But their lifetimes are measured in millions of years, and in the few decades given to us we hardly notice any change. A volcano on the other hand lives at a much faster pace, more in tune with our own. It is never quite the same, its plume of smoke changing in color and height, its craters resting or bubbling up, and fingers of fire darting down from newly opened fissures. One can see its shape change in the space of a few years, sometimes in a matter of days or even hours. The life of a volcano is one of the most direct manifestations of the earth's life. Watching a volcano is like feeling the pulse of the planet, because volcanic activity is the most palpable manifestation of the awesome forces that rupture the crust, so that molten rocks erupt from the mantle below, renewing ocean floors, moving continents, causing earthquakes, and changing the face of the planet and the living beings that inhabit it.

All this we shall now see better and better because with this key we shall finally follow, in broad lines, the life and changes of our planet, we shall understand how the things we see were born, and we shall discover some of the many worlds that our world has been in the past.

CONTINENTS OF THE PAST
From what we have seen it does not seem that the continents are apt to rise and sink easily, but rather to move, collide, join, and break away again, taking on new shapes and new positions on the earth's surface and forming new mountains and new seas.

As Wegener had suggested, around 200 million years ago all the continents were united in a single land mass, Pangaea, that no later than 150 million years ago broke up into the present continents. In effect Pangaea itself was made up of two great blocks (figure 56a) that had joined only about 100 million years earlier (figure 56c). One of the two comprised the southern lands—South America, Africa, Antarctica, Ceylon, Madagascar, peninsular India, and Australia—and was named Gondwanaland. The other consisted of North America, Greenland, Europe, and Asia and is known as Laurasia. In even earlier times the history of the two blocks is somewhat uncertain. Laurasia may have formed from minor land masses, while Gondwanaland seems to have existed from a very remote past.

Studying the Gondwanaland of 200 million years ago one finds that the oldest mountain chains, which began to form 3 billion years earlier, run parallel to one another. This could not have occurred by chance on drifting continents that had later joined, but it is entirely natural if the ranges formed when all those lands constituted a single block. In addition, the oldest ranges were in the center of Gondwanaland, while to the south rose younger ranges, also in parallel ramparts. All this argues for the great antiquity of Gondwanaland as a whole. The fact that each new mountain chain rose parallel to the previous ones also indicates that the motions of the mantle that were generating them occurred in the same direction for at least 3 billion years.

At this point, going back to H. Frey's theory of the formation of ocean basins, it seems quite likely that, as it cooled, the earth was entirely covered with a thin uniform crust, and that Gondwanaland was all that remained of that crust after the asteroids had crashed through it, creating the first ocean basins. It is not clear that plate tectonics

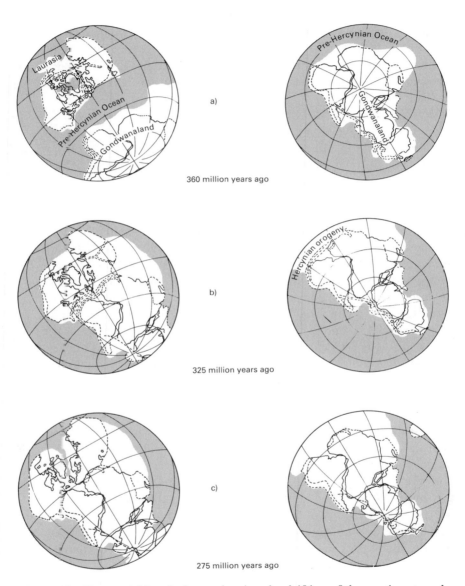

Figure 56   Terrestrial hemispheres showing the drifting of the continents and the poles between 360 and 25 million years ago.

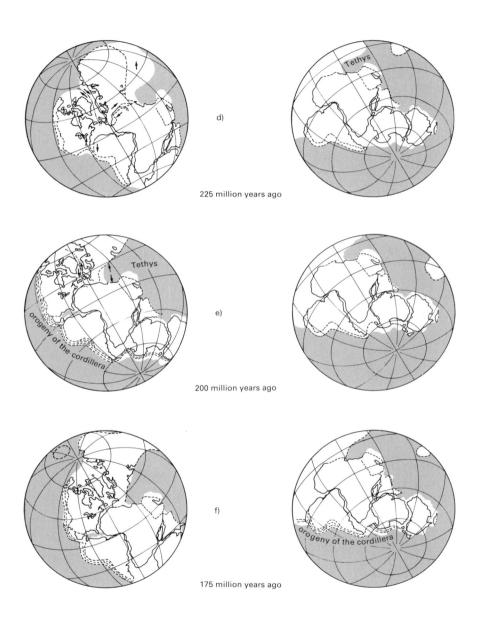

d)

225 million years ago

e)

200 million years ago

f)

175 million years ago

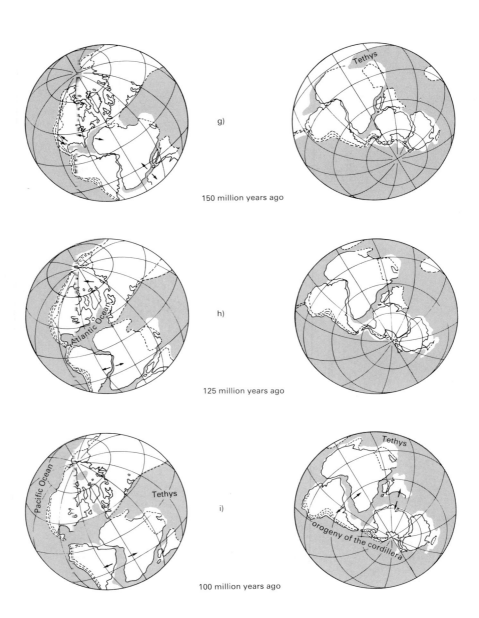

g)

150 million years ago

h)

125 million years ago

i)

100 million years ago

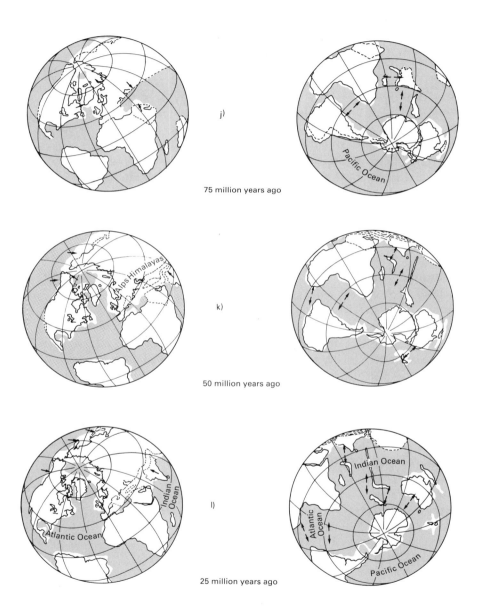

j)

75 million years ago

k)

50 million years ago

l)

25 million years ago

would have acted on a crust that covered the entire planet uniformly. By crushing a good deal of the surface, the asteroids carved out ocean basins, caused them to be covered by basaltic lava, divided the earth into continents and oceans, and, having caused these imbalances, started the drift process that is still going on. To this process we owe the variety of mountains, islands, volcanoes, seas, rivers, and lakes that make up the modern landscape.

Quite likely the asteroids fell on the Northern Hemisphere and only large fragments of the original crust survived there. After various recombinations they eventually formed Laurasia. To the south, most of the original crust was left in Gondwanaland, which remained intact until long after it joined Laurasia to form Pangaea; afterward it broke apart to form the present continents.

A large body of geologic, paleontological, and geographic data has permitted a fairly accurate reconstruction of the continents up to 200 million years ago. It is also possible to go back 500 million years, long before Pangaea formed, but with far less accuracy because the ocean floors of those ages no longer exist. In addition, the numerous paleomagnetic measurements made in recent years on the various continents have enabled E. Irving to pinpoint the positions of the magnetic pole during the past 375 million years. Assuming that the geographic pole has never deviated from the magnetic pole by more than 6–7°, he has drawn the shapes of the continental blocks and their positions on the globe starting from that time. This panoramic view shows vividly how the continents have changed and moved around in the past 360 million years, passing through polar zones, tropical zones, etc. (figure 56). Thus we see that *Lystrosaurus* could very well have lived in Antarctica 200 million years ago, since that continent was attached to Africa and India and was closer to the tropic than to the pole.

CATASTROPHES AND RENEWALS

THE FORESTS OF THE CARBONIFEROUS

We resume our journey 350 million years ago when plants and arthropods had spread over the continents and the first amphibians were commuting between land and water. Starting then, and for many tens of millions of years to come, in many areas of the earth there developed huge forests that, having later become fossilized as coal, gave the name Carboniferous (coal-bearing) to the entire period, which lasted 70 million years. This development occurred essentially in the Northern Hemisphere. At that time the equator passed through Scotland and most of Laurasia was in a tropical zone (figure 56b). In addition, a great deal of the land was flat and swampy, just a little above sea level. These favorable conditions resulted in a veritable explosion of plant life, which already counted almost 100 million years of evolution.

Entering a forest of the Carboniferous, we find it very different from those we know. It is certainly more similar to a jungle than to a forest, but is even more intricate. Giant, strange-looking trees, such as lepidodendrons, up to 40 m high, sigillarias and calamites, up to 30 m high, cordaites, from 30 to 40 m high, with trunks from 1 to 1.5 m in diameter, soared like columns toward the sky that they covered with their intricate canopies through which scant light would filter. Down below, the empty spaces between trunks were filled with pteridosperms, a sort of seed-producing fern that could live also on the mountains, and ferns of every type, including the Marattiales, of which a few species survive today in Malaysia. There were also sphenopsids, strange plants with jointed and grooved stems, like bamboo canes, which branched out with large and small stems that, depending on the available space, would expand at the base or rise with their knotty trunks to the top of the forest and only there would widen into enormous tufts.

This lush vegetation formed an almost impenetrable jungle. With swamps and bogs everywhere, the air was heavy with humidity and the smell of decay. The mantle of greenery stretched as far as the eye could see, but it was a monotonous landscape in shades of green, unrelieved by any splash of color because flowers did not yet exist. Silence reigned everywhere—unlike today's jungles, the forest of the Carboniferous was not populated by parrots, monkeys, or birds that cry or chirp.

Naturally there were other animals. Besides fish, various types of labyrinthodont amphibians populated the swamps. Small reptiles plied between land and water, and there were arthropods almost everywhere —giant cockroaches and a variety of spiders, scorpions, and millipedes, which had left the water in the course of the preceding 100 million years. The warm, humid, and shady environment also attracted the mollusks, such as snaillike gastropods of marine origin, as well as strange, multilegged animals up to 1.5 m long that apparently lived between the sea and the coastal underbrush and became extinct at the end of the Carboniferous.

But the veritable explosion in the animal kingdom of this period was that of the insects. Among them, some endowed with wings were the first terrestrial beings to fly and hover in space above the soil. Among the trees of the Carboniferous flew dragonflies apparently similar to the modern ones but much larger; some had a wingspread of 70 cm, while today the largest tropical species do not exceed 20 cm.

Once in a while one of the colossal trees of the forest toppled over and rotted, leaving the stump standing upright. Hollowed out by rain and humidity, the stumps eventually filled with mud and became deadly traps for small animals, some of which were unfortunate enough to fall in and suffocate. These distant tragedies have come to light with the discovery of their fossil remains still preserved inside the logs.

This was the forest of the Carboniferous—a strange environment, at the same time hospitable and dangerous, so different from similar ones known to us that even the reconstruction we can make today can give us only a pale idea of it.

THE GREAT GLACIAL ERA

Even as these forests flourished in the tropical climate of Laurasia, the earth entered a glacial era that lasted about 120 million years and like all glacial eras occurred in waves of deepening cold. At the same time the South Pole shifted from what is now southern Africa to a point in the ocean immediately south of Antarctica (figure 56). Thus the great icecap that formed around the pole at each new glaciation gradually moved over Gondwanaland in a southeasterly direction. Figure 57 shows the southern icecap 230 million years ago in one of the periods of greatest expansion, just before the end of the glacial era. Also shown is the path of the South Pole between 450 and 230 million years ago. Concurrently the North Pole shifted over the Pacific ocean in a diametrically opposite direction.

This glacial era, with its waves of formation and dissolution of ice around the South Pole, systematically lowered and raised the general level of the sea, from which it subtracted water to form glaciers in the mountain areas and particularly on the polar icecaps. According to a plausible hypothesis, when the icecaps formed and sea levels fell, many

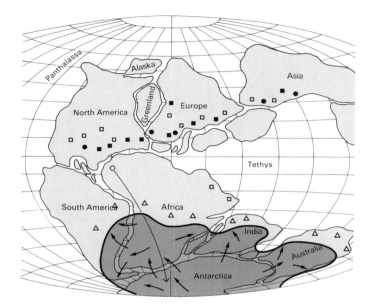

Figure 57   Rough outline of the land masses 230 million years ago. Note the extensive ice sheet and its expansion (thick arrows) and also the position of the South Pole 450 million years ago (circle) and its displacement over time (thin arrow extending from circle).

lands emerged and the great forests rose in Laurasia. Conversely, when the ice melted and the sea rose, lands and forests were submerged by water and layers of muddy sediments. Trapped in the mud without enough oxygen for rot to set in, trees and underbrush gradually turned into coal, the main nonvolatile component of wood. This is how the immense forests of the Northern Hemisphere—the likes of which had never been seen before and will never be seen again—produced the coal deposits we mine today. When we burn coal to warm our homes the heat we release is the heat of the sun stored by plants more than 300 million years ago. As glacial and interglacial periods alternated, forest grew on top of previous forest and each in turn became compacted into a layer of coal. Thus coal seams alternate with layers of soil. Having exhausted the upper seams formed by the last forests, we now have to dig deeper and deeper to search for the coal produced by older forests, all the way down to the first ones that rose about 340 million years ago.

As the great glacial era neared its height it brought about the downfall of the vast coal-age forests. There was no ice in Laurasia, but with so much atmospheric humidity trapped in the huge polar icecap in the south, the climate turned dry. The draught worsened during the Permian (between 280 and 225 million years ago) and vast stretches of sea dried up, leaving deposits of salt and chalk where no new forests could grow. In the first half of the Permian, lepidodendrons and sigillarias died out, and by the close of the period the pteridosperms had also disappeared. They were replaced by araucaria-type conifers, especially of the *Voltzia* genus, and ginkgoids, whose only descendant, the ginkgo tree, can reach a height of 25 m and is a native of Cekiang, China.

In Gondwanaland plant evolution took a different turn. Beginning in the Carboniferous there were recurrent and worsening glaciations, but after the first glaciers retreated vegetation returned with forms capable of surviving in the harsher climate. Thus there developed a distinctive flora consisting of plants like ferns, club mosses, and horsetails as well as a large number of primitive conifers, cordaites, and seed ferns. It was just a type of seed fern, *Glossopteris*, that left fossil remains scattered all over the southern continents, so widely separated today. This fact was one of the first arguments advanced by Wegener to demonstrate that all those continents once had been joined in a single land mass, Gondwanaland.

ANIMAL LIFE FROM THE CARBONIFEROUS TO THE PERMIAN

The amphibians, which had appeared in the Devonian almost 400 million years ago, branched out, evolved, and flourished during the Carboniferous and Permian over a span of 120 million years. In the following 40 million years they declined, and eventually only a few species survived. We shall soon watch their decline. But first we shall see how from these, during their ascending phase, a branch detached that would inherit the earth for about 150 million years, that still exists, and that would be one of the main causes of the end of the amphibians —the reptiles (figure 58).

The first reptile, though primitive, is found around the middle of the Carboniferous. It was *Hylonomous*, a sort of lizard, stocky, about 45 cm long, with four short legs and long feet, which descended from the amphibians, no one knows when and how.

With the reptiles there appeared the amniote egg, which was the cause of the reptiles' absolute supremacy over most of the other animals for many millions of years. This type of egg was different from the previous ones in that it did not need water like fish and amphibian eggs. In it, as anyone can see by breaking and examining a chicken egg, which is of the same type, the embryo that will become the new animal is wrapped in a membrane, the amnion; another substance is present, the yolk, to which the embryo is connected and from which it will derive nutrients until the birth of the animal. The amnion also contains a liquid that isolates and protects the embryo as the water of seas and lakes did for its predecessors. The whole thing is enveloped by a solid but porous shell through which the fetus can breathe.

This innovation permitted a spectacular expansion of terrestrial life. Since it provided the aquatic environment the embryo needed, the amniote egg could be laid anywhere on land, even in the most arid places. Furthermore, water was no longer needed to fertilize the egg. In fact, reptilian eggs were inseminated by copulation inside the female's body and not by simultaneous dispersion in the water of the two gametes (masculine and feminine), which is the way frogs still reproduce. Thus emancipated from the water, the reptiles acquired an enormous freedom and went on to occupy every possible environment—grasslands, dry forests of conifers, deserts, swamps, rivers—that is, the whole land. Operating in such diverse environments, mutations and natural selection generated a large number of species, so that during the Permian the reptiles, still of secondary importance at the end of the Carboniferous, had already become the dominant land vertebrates.

The main branch of reptiles, the cotylosaurs or stem reptiles, appeared during the Carboniferous (figure 58). There were both herbivorous and carnivorous species—a significant distinction, considering that up to that time all four-legged animals had been exclusively flesh eaters. Had there been only carnivores, as they spread over the earth the reptiles would have soon destroyed one another. But from then on the great majority of the animal population, 80–85%, would be herbivorous. This seems to be the optimal proportion for providing sufficient food for the carnivores while keeping the herbivores in check so that they will not devastate the vegetation. Thus the low percentage of carnivores ensured their survival, because had they been too numerous,

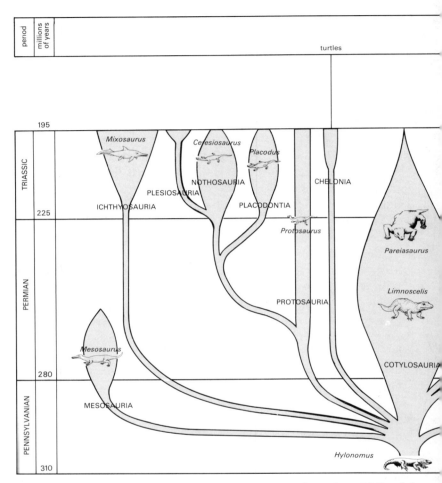

Figure 58   The first reptiles and their descendants from the middle of the Carboniferous. Some of their living descendants are shown in the strip at the top. [From *Il pianeta dell'uomo*, vol. VI, Mondadori; illustration by A. Fedini, Milan]

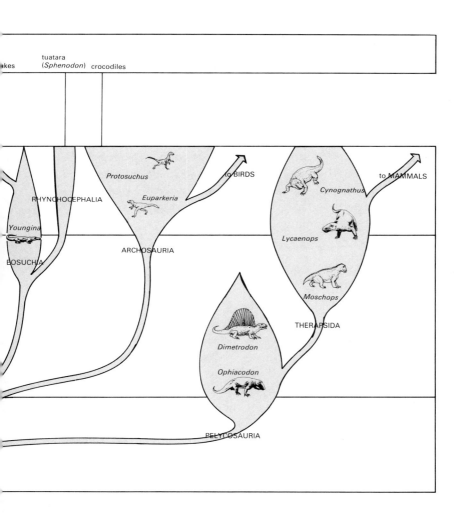

tuatara
(*Sphenodon*)   crocodiles

kes

RHYNCHOCEPHALIA

Protosuchus

Euparkeria

to BIRDS

Cynognathus

to MAMMALS

Youngina

Lycaenops

EOSUCHIA

ARCHOSAURIA

Moschops

Dimetrodon

THERAPSIDA

Ophiacodon

PELYCOSAURIA

after destroying the herbivores they would have eaten each other or starved to death.

A large number of Permian reptiles, living between 280 and 225 million years ago, were unearthed in the Karroo Desert at the southern end of Africa; others were discovered in the "red beds" of Texas, strange formations in the shape of columns or towers formed by erosion. In the Texas beds we find the pelycosaurs, which at some time in the Carboniferous had stemmed from the common cotylosaurian stock. To this order belonged strange-looking animals like *Edaphosaurus*, a herbivore, and *Dimetrodon* (the so-called Texas dragon), a carnivore. Both had developed vertebrae with long bony projections that were probably connected by a membrane, so that they appeared to be carrying a large fan on their backs. Presumably it was a device for regulating the temperature of these animals, cold-blooded like all reptilians. When hot, they would face the sun; when cold, they would stand sideways so that the membrane would receive the full sunlight, giving their bodies much more heat than could have been collected without that appendage.

The pelycosaurs became extinct before the end of the Permian, but not before giving rise to the therapsids, which would survive longer and in turn would give rise to the mammals. The therapsids included both carnivorous and herbivorous species, which lived together in the proportions we have noted were best suited to preserve the ecological balance: few meat eaters and lots of vegetarians. *Lystrosaurus* belonged to a family of therapsids. If you recall, this is the reptile that caused such a sensation among geologists when its fossil bones were found in Antarctica in 1969.

The mesosaurs are another group of reptiles that branched from the main cotylosaurian stock. Their remains have been found on opposite sides of the Atlantic—in western Africa, and in southeastern Brazil—and this was another of the facts used by Wegener as evidence of continental drift. A small river dweller, *Mesosaurus* appeared in the late Carboniferous and died out in the first half of the Permian, about 250 million years ago.

The full story of the reptiles unfolds in the three succeeding geologic periods, but we shall leave them now for a while in order to catch up with marine evolution and take a final look at the amphibians.

RENEWAL OF THE SEAS

After a period of decadence at the end of the Devonian over 350 million years ago, marine life revived with new types of plants and animals. During the Carboniferous the trilobites declined—they would become extinct in the Permian—while the graptolites and nautiloids nearly disappeared. (Of the latter, only a few species of *Nautilus* survive to this day.) They were replaced by new mollusks (ammonites), echinoderms, and crustaceans, as well as cartilaginous fish (sharks) and bony fish, which at the end of the Devonian had returned to the sea from the fresh waters. The ecological balance must have been based on the following food chain: Small crustaceans fed on algae and were prey to ammonites; the mollusks, in turn, were preyed upon by the bony fish, which themselves were food for the sharks. Thus the vertebrates were at the top of the food chain, as in today's seas.

At the end of the Permian, about 225 million years ago, marine fauna underwent a marked change, notably a general reduction in the abundance of life and the extinction of many characteristic Paleozoic groups. Shelled cephalopods were all but destroyed, though a surviving group, the ammonites, would subsequently give rise to a varied and immense population. The fish declined markedly, particularly those that had pioneered the move to the land—the Rhipidistia disappeared, while the crossopterygians and lungfishes came close to extinction and never recovered, though some species survived to this day. We do not know the reason for this radical change in marine life. Indeed, it is quite inexplicable, since this period was not marked by extensive glaciations or by climatic revolutions, at least on the surface. A cosmic phenomenon also can be excluded because no such catastrophic event is observed in the land faunas and vegetation that would have been even more vulnerable to it.

DECLINE OF THE AMPHIBIANS

The great renewal of marine life marked the end of the Permian, 225 million years ago. With it, the Paleozoic era, the era of the "old animals," came to an end and a new era began, the Mesozoic, the era of the "middle animals." During the first period of the new era, called Triassic, the climate was very mild; the southern icecap had disappeared and the average temperature again had risen.

In the Triassic, between 225 and 195 million years ago, many amphibians attained giant sizes; *Cyclotosaurus* was up to 4 m in length and
Mastodontosaurus over 5 m. But this period also marked the end of their
hegemony. Chased from the lands by the reptiles, the amphibians returned permanently to the sea, where they were safer. To remain in
the water forever, they ended by arresting the individual metamor-
phosis, giving up lungs and adulthood. In sum, their entire life would
take place away from dry land. Their evolution stopped, and only some
minor representatives (frogs, toads, caecilians, and salamanders) have
come down to us. But by then the amphibians had played their essential part in the evolution of life, because through them the vertebrates
had emerged from the water and spawned the reptiles. Now that the
latter dominated all the lands, the amphibians could withdraw.

THE NEW ANIMALS OF THE TRIASSIC
The Mesozoic era, which we have entered while following the downfall
of the amphibians, began with a proliferation of reptiles of all types
and sizes and culminated in the triumph of the largest animals that
ever walked the earth—the dinosaurs.

In the Triassic we find three groups of aquatic reptiles that had
branched out from the cotylosaurs many millions of years after the
mesosaurs and pelycosaurs had first appeared (see figure 58). Of the
three, only the ichthyosaurs lived in the high seas. The nothosaurs, 50
cm to 3 m long, and the placodonts, stockier and up to 4 m long, lived
along the coasts much like today's seals and walruses, which apparently
they also resembled in way of life and feeding habits. The nothosaurs
and placodonts became extinct at the end of the Triassic, but meanwhile the nothosaurs had spawned the plesiosaurs, which would attain
giant sizes in the next two periods.

In the Triassic we also find the eosuchians, a group of land reptiles
that had branched out from the cotylosaurs in the second half of the
Permian, about 240 million years ago. Some of them, including their
earliest ancestor, the small *Youngina*, reproduced until 65 million years
ago—a time that was to prove fatal for nearly all of the animals on the
planet. Nevertheless, they left progeny. The eosuchians almost certainly were the ancestors of lizards, geckos, monitors, iguanas, and
snakes. Back in the Permian the eosuchians had also spawned the rhyn-
chocephalians, whose last descendant, the tuatara, still survives in New

Zealand. But the richest and most interesting subclasses to stem from the cotylosaurs were undoubtedly the therapsids and the archosaurs, which produced, respectively, the mammals and the dinosaurs.

Let us begin with the archosaurs, or "ruling reptiles." The first archosaurs were the thecodonts ("having teeth set in sockets"), which appeared early in the Triassic and rapidly increased in number and variety, approximately between 230 and 220 million years ago (figure 59). One of the earliest was *Euparkeria*, a creature less than a meter long that resembled a crocodile more than a lizard but was neither. It was bipedal, and with its long sturdy legs it must have been a fast runner with which no four-legged animal could compete. The sturdy hind legs, the short front legs ending in paws, the long tail, and the shape of the head recall a carnivorous dinosaur, but it was not a dinosaur. Nor were any of the many other thecodonts that had rapidly evolved in a variety of shapes. *Euparkeria* was one of the primitive thecodonts of the suborder Pseudosuchia, almost all of them with the same characteristics, which in about 10 million years evolved into the first dinosaurs.

By the end of the Triassic, dinosaurs by the millions had spread over the entire planet—all kinds of dinosaurs, herbivorous and carnivorous, bipedal and quadrupedal, large and small, with some already showing a tendency toward gigantism.

THE GREAT AGE OF THE DINOSAURS

The spread of the dinosaurs marks the beginning of a new period 195 million years ago, the Jurassic, which lasted 60 million years. At that time the continents, populated by animals we have already met and many others less noticeable, were undergoing a radical transformation that did not result in a catastrophe only because the phenomenon occurred slowly and gradually.

As we have seen, about 300 million years ago the supercontinents Laurasia and Gondwanaland were joined in Pangaea. This great land mass, comprising all dry lands, did not behave like a single block for long, and perhaps it never really did. In any case, as E. Irving has recently deduced from a wealth of paleomagnetic measurements, about 225 million years ago North America moved north and Gondwanaland began to rotate counterclockwise, stopping only in the late Triassic. At that point North America reversed its motion and along with Gondwanaland began to rotate counterclockwise around Europe (figures

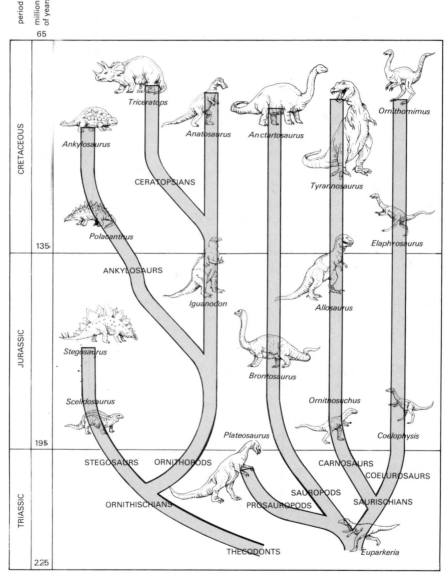

Figure 59　Progeny of the reptiles: the subclass *Archosauria* in its seven suborders; the suborder *Therapsida* subdivided into two groups, carnosaurs and coelurosaurs. [From *Il pianeta dell'uomo*, vol. VII, Mondadori; illustration by A. Fedini, Milan]

56d, 56e, and 56f). It was at that time, between 200 and 180 million years ago, that Pangaea began to break apart.

The separation must have started with eruptions of basaltic lava along the edges of the blocks that were to become the modern continents. Almost certainly, the basaltic Watchung range in New Jersey and the volcanic formations in the Connecticut Valley along the east coast of the United States, as well as the huge basalts of the Drakensberg in South Africa and the corresponding lava flows in Antarctica, are all evidence of these first fractures and separations. At the same time the Atlantic Ocean began to open up, and for at least 50 million years it remained a sort of inland sea. In the Pacific Ocean, meanwhile, a deep gulf called Tethys Sea formed between the southern coast of Eurasia and the coast of North Africa. Due to the northward pressure of Africa, 150 million years later this gulf was partially squeezed into the Mediterranean Sea. Approximately 140 million years ago Africa and South America began to separate, and by the end of the Cretaceous, 65 million years ago, all the modern continents had moved apart. However, their contours were not exactly the same as today's, the seas were different, and some land masses like peninsular India still were in the middle of the ocean (figure 56).

The separation of the continents had profound consequences on the land fauna. Take the dinosaurs, for example. At the beginning of their evolution all the lands were still joined and the dinosaurs could spread everywhere. Subsequently the various species evolved independently, with different forms, on continents separated by wide, deep seas. Exchanges between one continental block and another could occur only through intercontinental bridges such as the isthmus of Panama or Bering, but such exchanges would be limited and intermittent because the connecting passages were thin and sometimes obliterated by rising seas.

Let us see now what the land looked like at the beginning of the Jurassic, when a terrestrial animal was still free to walk over all the land.

The vegetation was dominated by the Bennettitales, plants with stocky stems and inconspicuous flowers. The conifers (araucarias, firs, and cypresses) also were abundant and widespread. The first sequoias were beginning to appear while some survivors from the past still lived, like cycads, large ferns, and the swamp-loving horsetails.

Besides forests and savannas there were steppes and deserts, and all these environments—some more, some less, depending on conditions

—were populated by hosts of arachnids, insects, and vertebrates. The amphibians survived only in small forms similar to those living today. With the exception of the crocodilians, the older reptilian groups had become or were becoming extinct. The thecodonts, unable to compete with the coelurosaurs and carnosaurs, their more advanced descendants, had disappeared during the Triassic. There remained, undisputed masters of the world, the dinosaurs (figures 59 and 60). They were entering their golden age, which was to last more than 130 million years.

Everyone is familiar with the giant dinosaurs. Reconstructed with some imagination on the basis of skeletal remains, they have appeared in hundreds of illustrations and in science fiction movies set in another time or in another world. Let us try to see them less and to get to know them better. One of the biggest was *Brontosaurus* ("thunder lizard"), which was over 20 m long. It had a massive body supported by four thick legs, a long neck, and a long tapering tail ending in a sort of whip, perhaps a defensive weapon against its smaller enemies. It weighed approximately 30 tons and probably spent much of its life in water, but it certainly walked on land, too, where the females laid their eggs. It lived in Laurasia. In North America, toward the end of the period, it acquired a companion, *Diplodocus*, which was nearly 30 m long but of more slender build, weighing not over 10 tons. But by far the largest of all dinosaurs, and the largest terrestrial animal of all times, was *Brachiosaurus*, which appeared at the end of the Jurassic. It lived in North America and in East Africa, where a skeleton 23 m long has been found. The live animal must have weighed 80 tons, and with its head it could reach a height of 12 m from the ground.

These animals, all herbivorous, were threatened and often killed by carnivorous dinosaurs. The fiercest one of the late Jurassic was *Allosaurus*, which was up to 10 m long and a biped; that is, it walked erect. With its powerful hind legs it could run fast, and by using its strong tail to balance itself it would lean forward, head outstretched, ready to grasp its prey in its huge mouth armed with long sharp teeth. Its short front limbs, similar to arms, had three-fingered paws armed with powerful claws. Naturally *Allosaurus*, which lived in North America, was not the only carnosaur of the time. Somewhat smaller, but equally fearsome, were *Megalosaurus* in Europe and *Ceratosaurus* in South America.

Herbivores and carnivores shared the same territory and normally

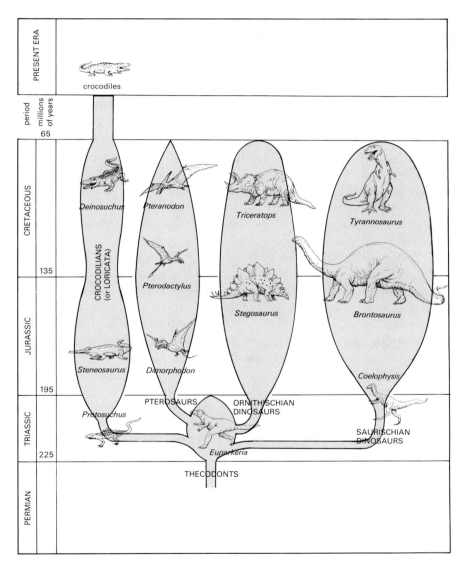

Figure 60   The dinosaurs: origin, development, and end from the Triassic to the close of the Cretaceous. [From *Il pianeta dell'uomo*, vol. VII, Mondadori; illustration by A. Fedini, Milan]

coexisted peacefully much as lions and gazelles do today. But every time a carnosaur was hungry the terrible, if natural, struggle for survival would be reenacted. Bound to eat or starve, the carnivore would attack the herbivore, which would seek safety in flight or in the water. We have impressive evidence of such life-and-death battles in the skeleton of a *Brontosaurus* whose bones bear the marks of an *Allosaurus*'s teeth, some of which broke off in the fight, leaving fragments nearby. Another hunting scene involving an *Allosaurus* chasing a *Brontosaurus* is no less dramatically suggested by the footprints they left on the ground. The battle between these titans must have been an awesome sight. One can almost imagine them trampling everything all around and making the very earth shake with their fury. If they had voices, something we do not know, they must have sounded more like rolls of thunder than the neighing of horses. Those of giant carnosaurs might have been terrifying enough to weaken their victims' will to resist.

Besides these colossal animals, there lived many small and medium-sized dinosaurs, such as various species of coelurosaurs, some tens of a centimeter to 3 m long. They were slender bipeds with hollow bones, like those of birds, and obviously built for speed. One can almost visualize them barely touching the ground as they chased their smaller prey, such as lizards, large insects, and mammals.

All the dinosaurs we have met thus far were saurischians.[32] The ornithischians, dinosaurs with birdlike pelvic girdles, also arose from the thecodonts in the Triassic, but they developed more slowly. They were all herbivorous and naturally many of them were heavily armored. Typical of these armored ornithischians is *Stegosaurus* (covered lizard), which was up to 6 m long and weighed about 2 tons. Its arched back rising steeply from the small head carried a double row of bony vertical plates. The sturdy tail was armed with two pairs of long bony spikes, which *Stegosaurus* probably used to defend its unprotected flanks. As it walked around, browsing through the vegetation and swinging its tail left and right, it must have done about as much damage as a modern tank. *Stegosaurus* lived in the second half of the Jurassic, between 160 and 135 million years ago, in North America and Europe. Similar animals have been found all over the world.

The stem reptiles, which at the beginning of the Triassic had spawned the dinosaurs, gave rise to other animals (mesosaurs, ichthyosaurs, plesiosaurs, etc.) that despite their names, given them when our knowledge was still limited, should not be confused with dinosaurs. Furthermore, all those animals were true reptiles, while the dinosaurs, though descended from reptilian stock, at a certain point seem to have lost at least one of the reptiles' fundamental characteristics. More about this later. Right now let us go back to those reptilian species that had returned to the sea (figure 58).

The ascent of marine reptiles had begun in the Permian, between 280 and 225 million years ago. We have already met one important group, the ichthyosaurs, which had branched off from the cotylosaurs no one knows when or how. They appeared in the Triassic, still in primitive forms, and reached their peak in the Jurassic. The ichthyosaurs lived in the open sea and were similar in appearance to modern dolphins; they were normally 1.5–4.5 m long, though some species attained a length of 12 m. Their shape is well documented because in addition to fossil bones we have found the imprints that their dead bodies left in the sand, which is now rock. We have also found something more poignant than mere imprints, vestiges of life and death at the time these two great events are closest to each other—the remains of female ichthyosaurs that died with their offspring during parturition. One female has been found with her babies still inside her, and another in the act of giving birth to a baby that evidently died with its mother and remained next to her, freezing the drama at the moment it occurred more than 140 million years ago. The discovery of their fossil remains has given us a glimpse into these distant tragedies, reminding us of the many others that have taken place ever since child bearing has involved suffering and danger for many females. That time had already arrived for female ichthyosaurs because these findings tell us that they were viviparous, like the mammals, or at least ovoviviparous. This is not surprising considering that they lived in the open sea and could not lay eggs on the shore, which would have been impossible anyway because of their conformation.

The plesiosaurs also were abundant in the Jurassic seas, but they would reach their peak in the next period. They descended from the nothosaurs of the Triassic. Though better adapted to marine life than their progenitors, they probably did not wander very far from the shores where they lay their eggs. Their main characteristics were four paddlelike limbs and a lizardlike head attached to a long neck that was very useful for plucking from the water the fish and cephalopods on which they fed.

Ichthyosaurs and plesiosaurs were not the only predators to prowl the seas; there were sharks, skates, and two types of marine crocodiles. Besides these large animals, there was a less conspicuous but very abundant fauna that fed on plants or smaller animals and included cephalopods, particularly belemnites and ammonites, corals, and bony fish—in sum, the vast minor flora and fauna that, except in the periods of widespread extinctions, had made the sea such a rich and varied environment for hundreds of millions of years.

THE CRETACEOUS

At the end of the Jurassic the world's climate became milder and the turn of the seasons more uniform. This produced an important innovation in marine life, which until then had consisted of animals that lived on the bottom or swam freely at various levels. The equable and mild climate favored the development of surface life, which already existed, including the tiny unicellular organisms both vegetable (coccoliths) and animal (Foraminifera) collectively known as plankton. These minute creatures had calcareous parts (disks and shells) that after death settled by the billions of billions on the bottom and in time formed thick deposits of chalk.

Starting from the same time, a good part of the land was gradually invaded by the oceans, so that around 85 million years ago shallow seas covered vast areas of the continents. About 20 million years after they had reached their maximum expansion the seas retreated, leaving great stretches of dry land. Meanwhile the shells of the creatures living on the surface for so many years had formed, in vast areas, thick layers of white material that gave a strange and desolate appearance to the lands from which the waters had just retreated. For this characteristic the period that goes from 135 to 65 million years ago was named Cretaceous. Some of these lands subsequently were uplifted by crustal movements we already know, and to this we owe the formation of a celebrated landmark, the White Cliffs of Dover.

The Cretaceous seas continued to harbor a varied and rapidly evolving flora and fauna. The plesiosaurs spread everywhere, attained giant sizes, and spawned new species, including short-neck forms. The most significant of the Cretaceous plesiosaurs, *Elasmosaurus*, may have reached 15 m in length, 9 of them in the neck alone, which contained 76 vertebrae. Presumably this animal used its long flexible neck to pluck its prey while keeping its body out of harm's way.

Unlike the plesiosaurs, the ichthyosaurs declined steadily. The marine crocodiles fared even worse and soon would become extinct. They were replaced by new reptiles, the large and ferocious mosasaurs, which descended from small and inconspicuous monitor-type lizards that early in the period had become adapted to life in the open sea. Thriving in their new ecological niche, where they replaced and perhaps dislodged the sea crocodiles, the mosasaurs multiplied and grew to large sizes, from 4–5 m up to 15 m in length. With their strong tails and paddlelike fins—a marine adaptation of their limbs—they could move fast, threading the waves with a snakelike motion. They had large mouths, sharp teeth, and an armored skin, and must have been very combative judging from the discovery of fossil bones with healed fractures, obviously battle scars.

Apart from animals that looked like creatures from another planet or from the world of mythology and fairy tales, the sea harbored some familiar fish: giant rays, sharks in growing numbers, and bony fish, which produced colossal forms never to be surpassed again.

The Cretaceous was not only the age of sea monsters and desolate white plains, which in fact were confined to just a few areas and appeared only at the close of the period when the shallow seas retreated; it also was the age of the beautification of the landscape. During this period, while the Bennettitales disappeared and the gingkoes, cycads, and horsetails diminished drastically, the angiosperms (flowering plants) appeared and evolved rapidly.

The early terrestrial landscape, stark and desolate, had just the colors of the rocks and the blue of the sea, tinted gray by the almost perennial cloud cover. Then the sun broke through the clouds, giving a new brilliance to rocks and water and coloring the blue sky with all the rich shades of twilight. Hundreds of millions of years later, when the great forests appeared, the landscape turned a restful green, but lacking flowers it was monotonous and rather somber, particularly when dominated by the dark green of the primitive conifers.

The spread of the angiosperms gave the environment a decidedly modern aspect. A host of new trees—oaks, sassafras, eucalyptus, plane trees, palms, magnolias, and walnuts, among others—spread over hill and dale; poplars grew in the plains and willows along the streams. At the turning of the seasons many trees made dazzling flowers, countless intertwined shrubs covered themselves with small multicolored flowers, and the prairies were constellated with tiny dots of all colors. A note of gaiety and a more gentle beauty descended over the planet like a fairy's mantle.

The sounds also changed because the insects multiplied, among them some buzzing pollinators, such as bees, necessary in order to pollinate the new flora. But this bucolic world still resounded with the heavy footfalls of the dinosaurs and perhaps the hunting cries of the giant carnosaurs.

There were all kinds of dinosaurs, both old and new. A famous new-comer was *Iguanodon* ("with iguana's teeth"), which appeared in Europe at the beginning of the Cretaceous and had relatives in North America, Africa, and Australia. It was a bipedal herbivorous ornithischian 9 m long and 5 m tall, rather queer looking and with a strange new attri-bute: instead of the third finger its upper limbs had an up-turning dag-gerlike claw that presumably was a defensive weapon.

But there was little that *Iguanodon* could have done against *Tyran-nosaurus* ("tyrant lizard"), the largest carnivore of all times. This bipedal saurischian of the late Cretaceous had a massive head about 1.5 m long and an overall length of 15 m. Its footsteps—as measured by the im-prints of its clawed, three-toed feet—spanned 4 m. Its huge jaws were armed with serrated, saberlike teeth of unequal length, which closed over its victims like an inescapable trap. This terrible creature was not unique. It belonged to a family of carnosaurs scattered over the conti-nents whose other members were somewhat smaller but equally fear-some. Nor were they the only carnivores around. The coelurosaurs continued to evolve in the Cretaceous, producing such specialized forms as *Ornithonimus*, which was 3 m long and looked like an ostrich because of its long neck.

The herbivorous ornithischians reacted to the growing threat of the carnivorous saurischians by improving and refining their means of de-fense, which did not remain passive, like the bony armor of *Stegosaurus*, but became also active. Some species acquired stereoscopic vision and an improved sense of smell, including the curious duck-billed hadro-saurs, whose jaws contained some 2,000 teeth that regrew as they wore down. Mummified specimens of these animals indicate that they had webbed feet and ate small evergreen branches, fruit, and seeds. Thus, while they may have been terrestrial in habits and diet, they were also adapted to water, where perhaps they took refuge when they "smelled" or saw danger approaching. The stegosaurs were replaced by the more heavily armored ankylosaurs, whose broad backs were solidly covered with a bony shield edged with stout spikes. There appeared also the ceratopsians, endowed with horns and bony collars. Best known is *Triceratops*, which lived in North America in late Cretaceous, was more than 10 m long, and weighed about 10 tons; despite its weight it apparently could gallop at 50 km/hour. Its head, 2.5 m long, ended with a powerful hooked beak and had a large bony collar to protect the neck. Three horns protruded from its skull, two long ones above the eyes and a shorter one above the nose. Thus armed, it could charge and kill, if not *Tyrannosaurus*, at least many of the other carnosaurs. Thus did the herbivores respond to the development of the carnivores that occurred during the Cretaceous.

More than a hundred million years after their birth the dinosaurs were still prospering and evolving into more sophisticated species. Though they dominated the scene, other animals populated the land. There were large lizards; giant crocodiles as much as 18 m long lurked in rivers and estuaries, and the first snakes were developing. The skies had changed and become more familiar because by now there were also birds.

All told, if we could journey back in time to the Precambrian or the Devonian, we would think we had landed on an alien planet, whereas by the end of the Cretaceous we would recognize our own earth, though strangely uncultivated and inhabited by creatures, all too real, that we had seen only in illustrations or science fiction movies.

LIFE IN THE AIR AND THE CRUCIAL PROBLEM OF THE DINOSAURS

Mention of the birds calls attention to a world we have not yet explored, the air. While the dinosaurs dominated the land in the Jurassic and Cretaceous, two groups of archosaurs had taken to the sky. They are collectively known as pterosaurs, or flying lizards. In effect the first ones were not much more than lizards with wings, but the pteranodons of the second half of the Cretaceous reached wingspans of 8 m and were the largest flying animals the earth has ever seen. They lived along the shore and probably fed on aquatic animals, much as sea gulls and pelicans do today. Judging from the number of fossil remains that have been found, they must have been very numerous.

Still in the Jurassic, while the saurians dominated both land and sky, the first feathered birds made their appearance. The oldest known is the famous *Archaeopteryx* (ancient wing), of the Jurassic. Only six fossil specimens of it have been found, and since 1861, when its remains yielded unmistakable evidence of feathers, this creature has been the object of intense study and debate, which have led to very important results. It was the size of a crow and it had a long tail, hind limbs endowed with three clawed fingers, and finally a saurian head with a toothed beak. Its skeleton was studied for a long time, and in 1937 J. H. Ostrom reached a startling conclusion: though feathered, this animal was a small coelurosaur, perhaps the smallest of all saurians. Further studies of its anatomy confirmed that it was a feathered coelurosaur and lacked the skeletal and muscular modifications needed for flight. How then do we explain the presence of feathers? They seem to have been just a thermal device to keep the body temperature constant. Does this mean that *Archaeopteryx* was a warm-blooded animal? And since it was a dinosaur, though small, does it mean that all dinosaurs were warm-blooded?

We have raised a question of great moment because it concerns one of the most important crossroads in the whole of animal history, namely, the crucial point at which evolution took a new route that would lead to mammals and mankind. In recent years numerous scholars, especially at Harvard and Yale, have addressed this very question. Let us review some aspects of it.

First of all we should recall what it means and involves to be warm- or cold-blooded. *Warm-blooded* animals (homeotherms) are those that keep a constant and relatively high body temperature in spite of external fluctuations. To do this they generate their own heat (endothermia). We call *cold-blooded* (ectotherms) those animals whose internal temperature varies with the environment and depends on an external source of heat, normally the sun. Of course this does not mean that a cold-blooded animal is always colder than a warm-blooded one. That occurs only when the environment is cold. Under these conditions the animal becomes sluggish and reduces its activity to a minimum. When heated by the sun, on the other hand, the animal may become even warmer than a warm-blooded one.

For every animal, food and respiration provide the energy necessary for all bodily functions and movement. The speed of the process depends on temperature, since chemical reactions are slower when the temperature falls below the optimal value for the enzymes to perform their function as catalysts. By releasing whatever little energy they have accumulated, cold-blooded animals are capable of brief periods of intense activity, but they cannot keep it up for long. In contrast, warm-blooded animals can be much more active, since they can sustain a high level of energy production for a long time. To do so they obviously need to eat much more. For example, a man must consume 40 times more food than a lizard of comparable size. Over 80% of the energy thus generated is used to keep the body's temperature constant. That is the price a warm-blooded animal has to pay. The great advantage, of course, is an active life instead of a long inactivity briefly interrupted by the exertion needed to flee from danger or search for food.

Let us see now on what evidence some modern scientists base their assertion that dinosaurs were warm-blooded animals. Let us begin with *Archaeopteryx*. The fact that this animal had feathers yet could not fly is very odd. From this observation Ostrom demonstrated *ad absurdum* that it was warm-blooded. Its anatomical structure, he reasoned, certainly categorizes it as a dinosaur, which, according to current opinion, is a reptile and hence cold-blooded. A reptile needs heat from the outside to sustain life, but feathers, by screening the body, would have prevented the animal from warming up. On the other hand, feathers are good thermal insulators. Hence feathers not used for flying could serve only one purpose, namely, to conserve the heat generated by the body itself. This implies that the animal produced heat internally, and hence one may conclude that *Archaeopteryx* was warm-blooded.

The other dinosaurs must have been warm-blooded too. According to Ostrom, the two-legged dinosaurs could not have walked upright without a high and stable body temperature. A lizard spends 90% of its life lying on the ground because it cannot stand very long even on four legs. The herbivorous dinosaurs needed to eat often because of their huge sizes and small mouths and, therefore, they had to stand and walk most of the time. The carnivores, beside standing upright on two legs, were frequently engaged in protracted hunts ending in long and furious battles with their prey. Just recall the chase of a brontosaur by an allosaur, recorded in their footprints, or the brontosaur's tail with dented bones and fragments of its assailant's teeth. In the Gobi Desert, moreover, fossil hunters have recently discovered the skeletal remains of a *Velociraptor* and a small *Protoceratops* still clutching each other in the fight that ended with the death of both. Hence our view of dinosaurs as ferocious fighters is supported by ample evidence.

If we could study the bodies of dinosaurs, we might find direct proof that they were warm-blooded. Unfortunately their soft parts have disintegrated, but bone fossils have yielded two important pieces of information. In some respects, having to do with animal physiology, dinosaurs' bones resemble mammalian bones and are quite different from the bones of reptiles and amphibians. Also, it has been found that the ratio between the length of the tibia and that of the femur is lower in slow moving animals (T/F $=$ 0.60 in elephants) than in fast-moving animals (T/F $=$ 0.92 in race horses; T/F $=$ 1.25 in deer). In the huge brontosaurs the tibia-to-femur ratio was the same as in elephants. In most hadrosaurs the ratio was higher than 0.80, and in many bipedal herbivorous dinosaurs it was over 1.00. This indicates that many of the dinosaurs were indeed the quick runners and fighters that we have imagined. And to be quick they had to be warm-blooded.

Warm blood would also explain the existence of very large dinosaurs. Take a dinosaur of modest size like a coelurosaur. Its body volume corresponds to the area of skin that envelops it. This area dissipates the heat generated by the body, and if the outside temperature is lower than that of the body, the animal feels cold. As the size increases, however, the body's volume and the heat generated increase as the cube, while the surface area increases as the square. (Think of the volumes and surface areas of spheres of increasing diameters.) Thus the heat lost by the larger animal will be proportionately less than the heat it produced. This means that with greater bulk it is easier to reach thermal equilibrium with the environment, and hence the animal needs to produce proportionately less heat.

For the very small dinosaurs like *Archaeopteryx* it was the other way around, and feathers or body hair must have developed as the best protection against the cold. In 1970 A. G. Sharov found in a fine-grained rock the fossil body of a small pterosaur so well preserved as to show that it was once covered with long thick fur, which also grew on the membrane of the wings. Consequently, it, too, must have been warm-blooded, and so, perhaps, was the entire order of pterosaurs. This case resembles that of *Archaeopteryx*, but this time what was found was not a reptile in a presumed bird, but a hairy covering on an animal assumed to be a reptile.

If this reasoning is correct, feathers did not sprout for the purpose of flying, although eventually they led to flight. The long "wing" feathers grown for thermal reasons by the smaller coelurosaurs (which did not have wings) may have helped them to chase their prey or to catch low-flying insects. In time, muscles suited for flight developed, wing bones strengthened, and the first true birds appeared.

When the pterosaurs died out at the end of the Cretaceous, mastery of the air passed to the birds that meanwhile had been evolving and would continue to evolve. Soon they constituted a class in their own right and became one of the most important groups of vertebrates of our era.

Hence the next question is, When and how did cold-blooded animals become warm-blooded? As I said, this event is one of the most important turning points in animal history because it marked the passage from a lazy, monotonous life to an active and varied one. It gave rise to dynamic animals of increasing skill and versatility that one day would produce man. And man would be able to create a history—a history that for the first time would tell of the transformation of the planet at the hands of one of its creatures, a history that is still unfolding and almost certainly will continue for an unforeseeable time.

Research in this field has been under way for a long time. Though some scientists as far back as the turn of the century held that the dinosaurs were warm-blooded, most of the results mentioned above were obtained in the 1970s and are not yet universally accepted.

The current view is that warm-bloodedness developed twice, in two distinct groups of animals, the therapsids and the thecodonts. The sequence of events is clearly displayed in the Karroo Basin in Africa, part of ancient Gondwanaland, where the remains of successive epochs are laid down in layers. There, in the second half of the Permian, about 250 million years ago, lived the first therapsids, which were cold-blooded. From mid-Triassic, about 210 million years ago, we can recognize a more advanced group, the theriodonts. Represented by *Cynognatus*, a doglike creature up to 1.8 m long, these animals had specialized teeth for better mastication (incisors, canines, and grinding cheeck teeth that would later become molars and premolars); they also had rib cages, indicative of good respiration and oxygenation. In sum, they had developed the two systems needed to utilize food better and to produce heat more efficiently. It also appears that they had vibrissae, the sensory bristles over the mouth that in cats are incorrectly called a moustache, which suggests that they may have had hair on the rest of their bodies. All this means that they were moving toward a more efficient production of energy and conservation of heat. In other words, they had become warm-blooded. The carnivores akin to the therapsids probably were moving in the same direction because their fossil remains show attempts at developing a secondary palate in order to breathe while eating.

Meanwhile, especially in Laurasia, the Permian archosaurs had spawned the thecodonts, some of which evolved into warm-blooded animals that in turn would give rise to the dinosaurs and pterosaurs. But how did these thecodonts acquire warm blood? We think that it all began when their progenitors developed strong hind legs and tails by living in the water. It seems that at that time they fed on small reptiles similar to the hippopotamus and with walrus tusks that browsed on the bottom of swamps. When the temperature rose in the Triassic, the swamps dried up, their prey died out, and the thecodonts were forced to search for new prey on land. Here, by virtue of their strong hind legs, they began to walk upright. By bringing their legs underneath their bodies they could acquire a longer stride and gain in speed, an obvious advantage in the hunt, but to do so they had to change their way of producing energy, which they needed not only to walk and run but also to stand up. In other words, they had to become warm-blooded. When or how the necessary anatomical and physiological adaptations occurred, we do not know. But this may be the way the first thecodonts pseudosuchians like *Euparkeria* developed, and from them the first dinosaurs descended and inherited warm blood and an active way of life. Once again, the consequences of this new vitality can be clearly seen in the Karroo Basin.

As we have noted, therapsids arose and developed essentially in Gondwanaland, while the thecodonts developed in Laurasia. Both kinds later spread all over Pangaea, but it was mainly the thecodonts that invaded the world. Thus we can see what happened by focusing on the Karroo, cradle of the therapsids. At a certain point the thecodonts arrived there and the therapsids began to decline. By the close of the Triassic, 195 million years ago, the therapsids were extinct. But even before the last therapsids had disappeared, the thecodonts too had met a tragic fate. Unable to compete with their more advanced dinosaur descendants, they were overwhelmed. Thus, by the end of the Triassic the dinosaurs had become the undisputed masters of the world.

While the thecodonts were spawning the dinosaurs that one day would destroy them, a new group of warm-blooded animals, the mammals, began to branch off from the therapsids. These first mammals took refuge in the only ecological niches that the dinosaurs could not occupy, namely, those suited to animals no bigger than 10 cm, too small even for the smallest dinosaur. Perhaps the mammals could have fully developed at that time, but in a world dominated by giants all they could do was to hide and wait. They waited 130 million years. But that long time was not wasted. They improved their dentition by, among other things, reducing the change of teeth to once in a lifetime; they developed self-regulating mechanisms such as panting and perspiring for times of emergency; and above all they managed to survive and reproduce. Perhaps one day better times would arrive, when they would be able to exploit fully the attributes they were acquiring: the advantage inherent in being mammals and the fact that most of them were placentals. In fact the earth eventually did become the world they seemed to be waiting for, a world for which they had been unconsciously preparing themselves and of which they would become the new and more evolved masters. But of the countless creatures that lived through the long years of waiting, none knew this.

THE EXTINCTION AT THE END OF THE CRETACEOUS

Sixty-five million years ago, at the height of their development and after ruling the world for over 160 million years, the dinosaurs became extinct. They were not the only creatures to disappear. With them went a large part of the old fauna that had evolved until the Cretaceous. Nearly all of the plankton on the surface, all of the free-swimming animals from ammonites to ichthyosaurs, and most of the bottom dwellers disappeared from the sea. The dinosaurs disappeared from the land and the pterosaurs from the sky. The disaster struck the vegetation, too, but to a lesser extent. The animals that inhabited rivers and lakes survived, along with a type of plankton that had developed at the end of the Cretaceous. On land, besides most of the plants, there remained essentially the small primitive mammals (insectivores and carnivores), the birds, and some reptiles. But the survivors, too, underwent a drastic reduction in number.

Nearly all of the species of the following era, the Tertiary, appeared immediately or sometime after the great extinction. Together with the few survivors of the disaster, they inherited a planet almost devoid of life, repopulated it, and through various adaptations evolved into the modern plant and animal species.

Many theories have been proposed to account for the mysterious biological disaster of 65 million years ago. None is acceptable unless it simultaneously explains three circumstances that appear certain,[33] namely,

1. that the event involved most of the living beings of the time; according to D. A. Russell 50–75% of the then-current species became extinct;
2. that it occurred in a very short time; a recent study of microscopic plankton fossils has demonstrated that plankton disappeared abruptly at the end of the Cretaceous; and
3. that it was a worldwide phenomenon.

Since there is evidence that at that time the earth was leading an otherwise uneventful life, with no glaciations, no variation in seawater temperature, and no violent crustal movement or volcanic activity, one is forced to conclude that the cause of the catastrophe should be sought in the sky rather than on the earth.

One theory that has gained widespread acceptance was advanced by W. A. Alvarez and his collaborators, including his son Walter, a geologist, on the basis of their studies of the rock layer at the boundary between the Cretaceous and the Tertiary. First studied near Gubbio, Italy, this layer clearly shows the sudden disappearance of plankton fossils and an equally sudden increase in the amount of iridium. Based on these findings and many other considerations, Alvarez and his co-workers concluded that the great extinction was caused by the fall of an asteroid about 10 km in diameter whose impact produced an enormous amount of dust, dispersal of which in the atmosphere obscured the sun and halted photosynthesis. The suppression of the first link in the food chain, the vegetable one, caused the extinction of all the beings that depended on it.

Several objections can be raised to this theory. Perhaps the most serious involves the effects of the dust envelope surrounding the earth. It would have certainly cut off the sunlight, prevented photosynthesis, and caused at first a drop in the temperature. But then the same dust layer would have trapped the heat of the sun in the underlying atmosphere, causing air temperature to rise markedly, perhaps hundreds of degrees; as a result, all life on earth would have ended.[34]

To me, it seems more realistic to link the mass extinction of the Cretaceous to an explosive event that occurred in space at about the same time at a distance of 880 light-years from the earth. It seems very unlikely that the coincidence in the dates is mere chance. Furthermore, if the recent data on x- and γ-ray emission from supernovae are correct, the radiation reaching the earth would have been lethal.[35]

Thus, perhaps, 65 million years ago there occurred a cataclysmic event in space near the earth that nearly obliterated life on the planet. Yet, from this tremendous explosion new stars were born that today shine in our sky and, quite possibly, new worlds that might be going through the same prelife phases the earth experienced 4 billion years ago. This one event may have brought both death and new life. And even on our devastated planet, though marking the end of an era, it lay the foundations of a renaissance that one day would lead to mankind.

THE NEW LIFE

When the catastrophe was over, the mammals, decimated but not wiped out, emerged from their burrows and faced a world to repopulate. The creatures that had ruled the earth for 170 million years were gone, along with many other species. There still were crocodiles in the swamps, snakes similar to boas and pythons in the forests, large amphibian lizards more than 2 m long, many families of birds, insects, and several kinds of arthropods. But the noisy, exuberant croud that had stomped through the earth for millions of years had disappeared.

The small mammals that now began to live in the open without fear certainly did not make much noise in this new world, but little by little they grew larger, evolved into specialized forms, and in the end spread over the face of the earth in countless numbers distributed among thousands of species.

ORIGIN AND EVOLUTION OF THE MAMMALS

As we have seen, the mammals originated 195 million years ago from the therapsids, the terrestrial reptiles that had ruled the world between the end of the Permian and the beginning of the Triassic. We believe that the therapsids were cold-blooded animals that became warm-blooded in the course of their evolution. They passed on this characteristic to the first mammals, which had body hair and a specialized dentition that by favoring the assimilation of food permitted a higher production of energy.

Very little is known about those first mammals. The fossil record has yielded only a few bones, many teeth, and, fortunately, the remains of a transitional animal, a therapsid living about 200 million years ago and not over 25 cm long, whose skull had a primary reptilian articulation and a secondary mammalian one. The few remains we have found show that the early mammals were tiny creatures not more than 20 cm long that survived in a world of giants and predators by spending most of their lives in crevices, burrows, and tree hollows. These first mammals (insectivores, carnivores, and rodents) have been classified on the basis of dental characteristics.

By about 130 million years ago the mammals had begun to separate into the first marsupials, similar to the modern North American opossum, and the first placental forms, akin to our hedgehogs and shrews. The refinement of the mechanisms that regulate body temperature, accomplished during the 100 million years of semiobscurity, gave the mammals a new vitality that certainly would help to make them the new masters of the world. But this was not the main attribute they acquired to reach this position.

As we have seen, the development of the amniote egg enabled first the reptiles and then the saurians to spread all over the world. Once again the most significant change involved the egg. Modern mammals are divided into three subclasses: placentals, marsupials, and monotremes. In placental mammals the fertilized egg is not laid outside but remains inside the mother's body attached to the wall of a new organ, the uterus. Here the embryo developing within the egg is connected to a temporary organ, the placenta, that enables it to draw nourishment from the mother until it is complete and sufficiently developed. At this point it is born, emerging from the mother, and the placenta is discarded, but the newborn is not yet independent because it must complete its growth. Until it is able to feed itself, it is fed by the mother's milk by means of another new organ, the system of mammary glands after which the entire class is named.

The marsupials also bear live young, but the young's brief period of gestation does not allow for adequate development. The immature infants crawl into a pouch, or marsupium, where they feed from the mother's nipples and go through a second phase of development. Thus they achieve most of their growth by breast feeding. The largest kangaroos, for example, 2 m tall when adult, are only 2 cm long at birth.

The monotremes still have many reptilian characteristics and in a sense represent an intermediate stage—they lay eggs but nurse their young.

Of the three, the placentals had the greatest potential from the very beginning; today they comprise 94% of all mammals. They had two great advantages over all other animals. First of all that of developing the body of the young more or less depending on the length of the gestation. This possibility has benefited mostly the brain, which is already so large in the fetus as to give the impression of a monstrous head on a slight body. This disproportion remains after birth, and in human beings it disappears only around 20 years of age. A larger brain implies greater intelligence. Moreover, in the more advanced mammals the increase in brain size has been accompanied by another important development: that of the cerebral cortex, or gray matter, which by means of complex convolutions covers an area much larger than the skull itself. The cerebral cortex is the site of the faculties of intelligence, learning, memory, and sensory coordination. In placental mammals, at birth the cerebral cortex is ready to perform its functions and can be gradually activated by the mother during the period of nursing. Thus when this period is over the individual starts from a step that its progenitors had reached through centuries of experience. Contrary to common belief, a kitten does not eat mice instinctively but is taught to do so by its mother. The same is true of lion cubs and in general of all the placental mammals. This is the second advantage.

All this explains why the Cenozoic era, lasting from 65 million years ago to the present, saw the rise of the mammals, especially the placentals, and the emergence of one particular species, man, that at least on our planet would attain a position of supremacy over the living world.

The growth of the mammals, which I shall sketch only in broad lines and in its essential points, began in the Paleocene, between 65 and 55 million years ago, with the development of primitive hedgehogs, rodents, and the progenitors of hares and rabbits. There were some orders that would die out before 40 million years ago and the primates had already appeared. The Condylarthra, an order that existed since the Cretaceous, also developed between 65 and 40 million years ago. Among them appeared animals similar to bears and, around 50 million years ago, animals similar to dogs and wolves. In that period they produced the largest of the carnivorous land mammals—found in Mongolia, *Andrewsarchus* was a wolflike creature 4 m long and 2 m tall with the teeth of a carnosaur. The condylarths became extinct 40 million years ago, but back in the Paleocene (between 65 and 55 million years ago) they had spawned various orders of ungulate mammals, two of which began to diversify and multiply in the Eocene, between 50 and 40 million years ago. They were the artiodactyls, which have come down to us with the pig, hippopotamus, and deer, and the perissodactyls, of which survive the horse, tapir, and rhinoceros, among others.[36]

Meanwhile, the mammals had taken to the air with various forms of bats, which had developed back in the Paleocene, the time of life's recovery. In the same epoch the mammals had also begun to venture into the sea, where they would evolve into pinnipeds (seals), sirenians, and cetaceans, large carnivores from the beginning. The most impressive cetacean of all times was *Basilosaurus,* which lived early in the Eocene in the American and European seas. A large carnivore over 20 m long, it looked more like a snake than a whale and had a crocodilian head whose jaws were armed with serrated triangular teeth. Naturally the Eocene saw also minor mammals, such as rodents of every type, small marsupials, and some primates that lived in the trees.

The 'dawn horse," *Eohippus,* 25–50 cm high at the shoulder, appeared about 55 million years ago among the minor mammals of North America. By 3 million years ago it had evolved into the modern horse, *Equus,* which migrated from North America and populated the world except Australia. Between 45 and 35 million years ago, in what is now Egypt, there lived an animal similar to a large pig with a longish body and a special dentition that comprised numerous molars and four small tusks, two above and two below. This animal, called *Moeritherium,* was the progenitor of all the proboscideans living and extinct, including mastodonts, mammoths, and elephants. While evolving in height, shape, and number of tusks, the mastodonts spread over Africa and into central Asia. From there, between 25 and 3 million years ago, they invaded Europe, populated Asia, and finally by crossing the Bering land bridge, when it was not interrupted by the sea, made their way into North America. In the Pleistocene, less than 2 million years ago, the last branches of mastodonts crossed into South America over the Isthmus of Panama. In Africa, Asia, and Europe they became extinct about 1.5 million years ago, while in the two Americas they survived until a few thousand years ago.

Meanwhile, about 25 million years ago, there appeared in Africa an intermediate animal, *Stegolophodon,* which subsequently migrated to Europe and Asia as far as the East Indies, where it lived on until a few thousand years ago. From this animal, about 3 million years ago, descended the first real elephant, *Stegodon,* which was more advanced than the old proboscideans, now in full decline, and soon took over their habitats. In the following millennia *Stegodon* gave rise, through intermediate forms, to the African elephant, the Indian elephant, and the mammoth. The last had long tusks that protruded downward and then curled upward, a long trunk, and small ears. It was very tall (some European specimens attained a height of 4.5 m) and was covered with a shaggy, reddish-brown fur that along with a layer of fat under the skin protected it from the cold. The details of its body are known to us because some "woolly mammoths" have been found well preserved in ice or tar. Prehistoric man was personally acquainted with the mammoths and painted them on the walls of his caves. At the end of the Paleocene, about 10,000 years ago, the ancient elephants, including the mammoths, died out and only the modern elephants survived. At this point we have arrived at the present time. Let us return to the mammals of about 50 million years ago that we had temporarily abandoned.

The rhinoceroses already existed at the beginning of the Eocene in three distinct families—the running rhinoceroses, the water rhinoceroses, similar to the hippopotamus, and the "true" rhinoceroses. The last ones attained very large sizes and about 30 million years ago produced the largest land mammal that ever lived, *Baluchitherium,* which stood 5.4 m high at the shoulder. All three types were hornless and lived in the Northern Hemisphere. The tapirs also developed in the northern lands, from which they have now disappeared; they moved to South America only about 3 million years ago.

From 40 to 25 million years ago there developed giant boars, animals similar to the pig and hippopotamus that lived in rivers and swamps, and primitive ruminants, commonly called "ruminant pigs." The last ones lived in North America, where they were more numerous than all the other mammals put together. Meanwhile, in North America appeared also the first camels, which then spread throughout the world, evolving into more slender forms. In North Africa, the same period saw the appearance of new primates much more advanced than those of the Eocene. The vast herds of herbivorous mammals were threatened by birds of prey and a large assortment of old and new carnivores, including the saber-toothed tigers. Dating back to the end of the Eocene, these fearsome predators continued to evolve through the Miocene and lived on until about 10,000 years ago.

Thus we come to the Miocene, a period that lasted from 25 to 7 million years ago and saw the original mammalian fauna begin to turn into the modern one. The giant boars reached their peak about 20 million years ago, but it was their last gasp; soon afterward they became extinct and only "true" boars were left. Meanwhile, about 10 million years ago the giraffes appeared in Europe and Asia and later spread to Africa. Deer, too, made their appearance in Asia and Africa, while the American branch of the family was evolving from primitive cervids. Horses, rhinoceroses, and tapirs continued to evolve and multiply.

In the Miocene seas already lived dolphins, porpoises, and primitive whales, and the seals had begun to leave the shores. Of course, there were also all the life forms that had survived the ecological disaster of the Cretaceous, and particularly the bony fish, which had been thriving since then. Land and sky harbored also multitudes of insects, which had been increasing in numbers and varieties from the time they had first appeared on the earth. The sky was also alive with an increasing number of birds. Starting in the Cretaceous with a few families, they did not seem to be affected by the catastrophe that occurred at the end of that period. In fact, 65 million years ago the number of bird families was 45, while at the end of the Paleocene, 10 million years later, they had risen to 84, and in recent times, despite numerous extinctions, the number was more than 130. Never was life on earth so full, so rich in forms, so abundant in number of living beings as in the Miocene. And it continued this way for a few million years more. But it was not to last to the present day.

During the Pliocene and most of the Pleistocene, from 7 million to 10,000 years ago, the mammals became increasingly similar to the modern ones. Of the families that lived in the Northern Hemisphere during the Pliocene, 80% still exist today. During that period there appeared new carnivores and rodents that have come down to the present. More generally, about 7 million years ago Europe and Asia began to be populated by a fauna strikingly similar to that of modern Africa, with herbivores such as bovids, deer, and giraffes in rapid evolution. Camels and pronghorns flourished in North America, and everywhere, alongside the local fauna, lived rhinoceros, mastodonts, bears, saber-toothed tigers, and, coming down to smaller and smaller dimensions, hordes of rodents and insectivores.

The vegetation was more or less that of the Cretaceous, except for the lycopsids, which had undergone a drastic reduction after the disaster of 65 million years ago, and the flowering plants, which had been constantly increasing.

Thus far our journey in time has taken us, though briefly, to the lands of the Northern Hemisphere and Africa. What was happening meanwhile on the other continents that had been moving apart since the breakup of Pangaea 200 million years ago?

THE MAMMALS IN SOUTH AMERICA AND AUSTRALIA

After the great extinction of 65 million years ago, the first surviving mammals began their ascent also in old Gondwanaland, which had been breaking apart for a long time. By then, Africa had separated, but South America, Antarctica, and Australia were still connected (see figure 56j). Some think that an isthmus may have linked South America to North America, but the connection was certainly broken in the Eocene, some 10 million years later. Between 55 and 40 million years ago Antarctica and Australia also separated (figures 56k and 56l), becoming island-worlds where life continued to evolve independently of that in the other continents.

Let us first look at South America. There, like everywhere else, lived the descendants of the primitive marsupials that along with the other mammals had appeared during the lifetime of the dinosaurs. In addition to pouched mammals, the original South American fauna included various types of ungulates and strange animals like armadillos, sloths, and anteaters that between 25 and 10 million years ago grew to large sizes. The armadillos also acquired a sturdy armor and offensive weapons such as a powerful mace at the end of the tail with which they could protect themselves from pack hunters.

Despite South America's isolation, the fossil record shows that evolution there produced animals analogous to those that were developing at the same time in the Northern Hemisphere. The reason, we believe, is that they lived in similar environments and performed similar functions.

Let us stop here a moment and consider this important phenomenon of an independent and parallel evolution, which we shall discuss again later with regard to extraterrestrial life. I shall give some examples, starting with South American ungulates. There were four orders, all extinct today, and all of them included animals similar to those on the other continents. Some, like *Nesodon* of the Eocene and *Toxodon* of the late Pliocene, though hornless, resembled the rhinoceros. Others, like *Theosodon* of the Miocene and *Macrauchenia* of the Pliocene, despite the latter's short trunk, had characteristics similar to the modern llamas and to the camel of North Africa. *Thoatherium* and *Diadiaphorus*, both of which appeared in the Miocene about 25 million years ago, looked very much like horses, especially in the feet, which soon turned from three toes into a hoof. Neither of these two animals lasted long: the first died out in mid-Miocene about 16 million years ago, and the second at the beginning of the Pliocene, 7 million years ago. Another animal, *Astrapotherium*, which lived from the Oligocene to the end of the Miocene, was nearly 3 m long, had a small trunk, and resembled a tapir. Finally, *Pyrotherium*, which lived from 50 to 25 million years ago, was the size of an elephant and like the latter had tusks and a good-sized trunk.

Many of the evolving marsupials, like the opossum, kept the characteristics inherited from their primitive forerunners of the Cretaceous. Others, instead, turned into flesh eaters. Luckily for the other animals, this must have occurred late, and perhaps slowly, since the most successful and most ferocious specimens flourished in the Miocene, between 25 and 7 million years ago. They included a sort of marsupial "wolf" and marsupial "felines," one of which was the size of a leopard and was saber toothed. Thus marsupial carnivores paralleled many of the placental orders in the north.

To these predators were added birds that had lost the ability to fly but made up for it in size, like the 1.5-m tall, sturdy-legged *Phororhacos,* which in the Miocene roamed the Patagonian plateau hunting small prey with its hooked beak. In the Pliocene and Pleistocene this bird was replaced by others of equal or larger size and no less ferocious.

Starting in the Oligocene about 30 million years ago, this fauna was infiltrated by minor mammals, such as rodents, monkeys, and raccoons, that were not natives of South America. They probably arrived there from the central American region aboard masses of vegetation and tree trunks, perhaps with the help of land bridges. Thereafter these mammals evolved isolated from the rest of the world, like all the other animals on the South American continent.

South American isolation came to an end about 3 million years ago when the Isthmus of Panama formed again. In wave after wave, hordes of placental mammals then came down from the north. First came insectivores, rabbits, hares, squirrels, and rats; they were soon followed by canids, bears, weasels, small felines, and large cats like the mountain lions and the saber-toothed tigers; then came the mastodonts, and, last, horses, camels, tapirs, llamas, and deer. This invasion had tragic consequences for most of the indigenous fauna. The marsupials were exterminated, including the flesh eaters that resisted longer but in the end had to give way to the placental carnivores from the north, more intelligent and versatile, which took over their ecological niches. The old hoofed mammals were supplanted by the new, with the exception of *Toxodon* and *Macrauchenia.* The armadillos and sloths survived, the former by growing in size and developing a more efficient armor, the latter by evolving into larger animals, the megatheres, some of which were bigger than elephants. Subsequently, sloths and armadillos migrated northward, followed by some minor mammals. Although this counterinvasion was much less imposing than the flow in the opposite direction, it served to enrich the North American fauna with species that still exist.

After separating from Antarctica, probably in the Eocene, Australia also became an island increasingly farther away from the other continents. There were no further contacts, and this resulted in the undisputed supremacy of the marsupials that had developed there since the Cretaceous or had emigrated from South America through Antarctica before the connections among the three continents were severed. Until less than three centuries ago Australian fauna—the result of over 40 million years of evolution—consisted of monotremes, some placental mammals, and, above all, marsupials. The monotremes, including the well-known platypus, are strange animals that seem to be made from pieces of other animals. The placentals were represented by some types of rodents, bats, and the doglike dingo. Rodents and bats must have arrived there by sea on carpets of leaves and tree trunks, just like the small placentals that had reached South America during the interruption of the Isthmus of Panama. The dingo, instead, almost certainly was brought there by the first humans to reach the continent. Neither the monotremes nor the few placentals disturbed the marsupials, which produced numerous, still extant species. Today, apart from the Tylacinus, which seems to have become extinct in the past few years, we find Kangaroos, bandicoots, koalas, wombats, the squirrellike phalangers, and among the smallest ones, marsupial "rats," "mice," and "moles."

Thus the evolution of placentals and marsupials can be summarized as follows. Born at the time of the dinosaurs but developed only after their demise, they both spread all over the world. The placentals, more intelligent by virtue of a more advanced brain, supplanted the marsupials almost everywhere; the latter flourished in South America, and from there, through Antarctica, they moved to Australia. When the northern placentals invaded South America by way of the Panama land bridge, the marsupials were almost completely destroyed. Australia, which had drifted away from the other continents and moved toward the equator, thus enjoying a milder climate, remained their only refuge, where they could evolve without competition from stronger adversaries.

In conclusion, the history of placentals and marsupials demonstrates once again that the evolution of life on earth is directed toward the ascendancy of the forms that are more intelligent, more complex, and more adaptable. Remember, too, that the placentals' way of reproducing is the most dangerous and many females die in childbirth, while reproduction involves very little danger for the marsupials and none at all for egg-laying animals. The fact that in spite of this evolution has occurred through the placentals is further proof that in the living world the sacrifice of the individual matters nothing vis-à-vis the improvement of the species.

THE NEW ICE AGE

The disappearance-reappearance of the Isthmus of Panama and the final division of the last three pieces of ancient Gondwanaland are only two instances of the great physical changes that the earth has undergone in the past 65 million years. During all this time the continents have continued to drift. Although separated by the Atlantic ocean, Europe and North America must have remained connected at their northern tips for a while longer. But by 50 million years ago the ocean had cut off Greenland, which became an island, and that link between North America and Eurasia was definitely severed, though a link remained on the Pacific side through the Bering straits, then dry.

Some 45 million years ago India collided with Asia after drifting northward for tens of millions of years and began to slip under the continent. It has continued to advance ever since, and from the beginning of the collision it has pushed northward as much as 2,000 km in relation to Eurasia. As a result of this collision, about 35 million years ago the mountain chain of the Himalayas began to rise.

Meanwhile, around 50 million years ago the eastern end of the Tethys Sea began to close. With the welding of ancient Palestine, Syria, and Turkey at one end and the subsequent joining of Africa and Spain at the other, it became a large salt lake, the Mediterranean. In the past 40 million years the pressure of the African Plate against the European formed first the Pyrenees, then the Alps, and finally, from about 7 million years ago, the Carpathian chain. In the two Americas, meanwhile, colliding plates built the Rockies in the north and the imposing cordilleras of the Andes in the south (see figure 55).

By causing the so-called eternal snows to accumulate on the higher elevations and by dividing the continents, mountain chains helped to bring about a change in the earth's climate. But this was one of the minor contributing factors. As the new ranges rose, the drift of the continents and the migration of the poles brought the South Pole to coincide with the middle of Antarctica, a continent of considerable average elevation. As a result, a thick icecap formed over Antarctica, which became a reservoir of cold for the whole planet. Later on, another icecap formed in our hemisphere due to the rearrangement of the northern continents. By 30 million years ago these lands were located in a circle around the North Pole, enclosing a sea nearly centered on the pole itself. Due to the scant exchange of water with the other oceans, the Arctic Sea became colder and colder, began to freeze, and eventually was covered over by a thick ice sheet—another large reservoir of cold.

All these large-scale geologic changes had a profound effect on the earth's climate, which at the beginning of the Cenozoic era was still uniformly mild and subtropical, that is, warm and humid. The change in the climate started about 30 million years ago. First there was a slight but continuous drop in the temperature. Then came small fluctuations that amplified the effects of Milankovič's seasons, causing large oscillations and cold waves, and the earth entered a new glacial era. Climatic zones were accentuated; the formation and spread of glaciers lowered the level of the seas, and new lands emerged, giving the climate a continental character with strong extremes of temperature.

This radical change in the climate, the evolution of the mammals into modern forms, and the increase and spread of the flowering plants made the earth the world we know. Our journey in time, which began with the formation of the sun and the solar system, has finally brought us back to our world, which we shall now follow in its recent vicissitudes.

As noted, the first signs of a change in the climate appeared around 30 million years ago when nuclei of ice began to form in Alaska and Antarctica. At first small, the ice sheets became increasingly more extended and stable, and finally spread to form vast and permanent icecaps. About 3 million years ago the climate became still harsher, and by the dawn of the Pleistocene a new glacial era had begun. As I mentioned earlier in the book, it is still going on and, like the past ones, is characterized by waves of intense cold separated by milder interglacial periods. The last glaciation reached its peak about 18,000 years ago and ended 10,000 years ago. Since then the earth has been in an interglacial period, but for the past 5,000 years the average temperature has been falling again.

The glaciations and warm spells of the past 2 million years have had a profound influence on animal life and evolution. Some species perished, many migrated several times to follow a favorable environment that was shifting over the globe, and others quickly adapted to the new climatic conditions. The cold waves produced the woolly mammoth, the shaggy-coated American mastodont of the Quaternary period, the woolly rhinoceros, and *Elasmotherium*, a mammal similar to a rhinoceros but with a single horn that lived in Europe and in Siberia. The cave bear spread from the Pyrenees to the Caspian Sea, while the last expansion of the glaciers drove the reindeer and the musk-ox south into Europe. These two have survived to the present, but unfortunately *Megaloceros* (the Irish elk) has become extinct. A large deer somewhat similar to the fallow deer, it is famous for having been endowed with the largest antlers ever known; in some specimens, living in Ireland, they stretched 3 m from tip to tip.

This splendid animal became extinct a little over 10,000 years ago during a general crisis that in a very short time brought about the extinction of numerous other mammals. Mammoths and mastodonts, megatheres and giant armadillos, saber-toothed tigers and cave bears, woolly rhinoceroses, beavers as big as bears, and large marsupials—they all vanished from the face of the earth. Minor mammals, including many types of rodents, also disappeared. Several European and North American mammals moved elsewhere; among those that left Europe were hyenas, leopards, lions, wild dogs, and hippopotamuses. In the two Americas several mammals failed to survive. In North America, where it had flourished for millions of years, even the horse died off. It returned to that continent only in the modern era when it was brought back from the Old World by the Europeans.

This sudden disappearance of a large number of species has not yet been explained. It seems the more strange in that it occurred at the end of a cold spell that those animals had survived. Perhaps they died of starvation, because the glaciations caused droughts in their natural habitats. Perhaps it all happened because meanwhile another mammal had developed and evolved that we have not met yet but had already existed for some time, an animal that knew those other animals, painted their likenesses on cave walls (because it had artistic talent), and hunted them to feed and defend itself, because it, too, was trying to survive. This mammal, so versatile, had also gone through and survived the first waves of the new glacial era and had already evolved to the point of being what it would look like today: man.

MAN

The history of man is rooted in epochs far more remote than those in which human beings began to look like you and me. Since Darwin's time it has been said that man has (or has not) descended from the apes. This is a shortsighted view. Our forerunners should not be sought among the early primates, for they were only our most recent ancestors. If we trace our lineage back in time, we find that our origins go back to the first placental mammals, and before them to the reptiles, and going further back to all that was preparing the next step. It is true that as we get close to our time we find "brothers" of our progenitors, more or less remote, that have evolved differently, but it is also true that today we would not exist, or we would be different, if even one link of that long chain were missing, from which, as it formed, many others also branched out. To understand the evolutionary history of the universe also means to be able to place our origin in the context of the origin of all things and all beings, known or unknown. It is what we are doing and shall continue to do in our journey in time. And now the time has come to discuss man himself and his origin within the vast class of mammals, which, I wish to stress once again, was already the result of a large number of evolutionary steps on the route that led to the first men.

As you recall, the first mammals appeared in the Triassic, about 200 million years ago, and separated into marsupials and placentals in mid-Jurassic, about 40 million years later. At the beginning of the Cretaceous, in the dinosaurs' heyday, two branches formed, one leading to the insectivores and the other to all the remaining orders. For most of the Cretaceous there were only the insectivores, but by the end of the period the order of the primates had branched off from them. Thus the primates were already in existence before the disaster of the Cretaceous-Tertiary boundary.

The link between insectivores and primates almost certainly was an ancestor of the tree shrew, an animal that lives in some region of the Far East and may not be too different from its progenitors of over 50 million years ago. It looked more like a tree shrew than a monkey, and unlike the other mammalian insectivores, which browse on the ground for food, it lived in trees. Because of its arboreal way of life, this ancient animal developed sight and hearing while its sense of smell decayed; these are characteristics found in both apes and humans. Lacking an opposable thumb, it climbed trees with the help of its claws and tail.

The development of an opposable thumb and the forward placement of the eyes were the major evolutionary steps accomplished by the primates, and again they were a consequence of their arboreal life. The first enabled them to grasp branches more securely, and this fact in turn led to a much greater mobility of the forelimbs, which soon became arms. By moving their eyes from the sides to the front of the face they acquired the stereoscopic vision necessary to distinguish a nearby branch from a faraway one and hence to jump from tree to tree. Depth perception already had been achieved by some of the last dinosaurs and later would be developed by the felines and some birds, such as the owls.

The evolution of the monkeys began approximately 50 million years ago with the so-called prosimians. The isolation of South America, which began at about the same time, had an effect also on the evolution of the monkeys. The New World, or platyrrhine, monkeys that developed there led and still lead a thoroughly arboreal life; they are rather small and have a flat nose and a prehensile tail. In the rest of the world, and particularly in Africa, the monkeys evolved into larger forms with a downturning nose and closer nostrils. Many of these Old World, or catarrhine, monkeys descended from the trees to live on the plains, and their tails either stopped being prehensile or disappeared altogether. It is within this family that the genus *Homo* developed. According to the latest research into the habitats of the hominoids, it appears that between 20 and 18 million years ago they lived in vast tropical forests, were tree dwellers, and fed on fruit. Between 15 and 12 million years ago they inhabited forested areas of both tropical and temperate zones and had already become partially terrestrial and omnivorous. A turning point in man's evolution was the transition to an upright stance with the body resting on two feet. Exactly when this occurred is not known. The oldest traces of bipedal gait were discovered in Tanzania in 1980 and date to 3.6–3.7 million years ago. Thus we know for a fact that at that time man walked erect, though he may have done so long before. In any case, the arboreal life initially common to all the primates made for a difficult gestation. This had the effect of reducing the number of offspring to one per birth and hence of prompting greater parental care of the young. Thus the primates gained an advantage over the other mammals, which already were in a privileged position with respect to egg-laying animals, which gave their offspring little or no parental care.

The great apes, such as orangutans, chimpanzees, and gorillas, at present are the most highly evolved. They resemble man (especially the chimpanzee) in anatomical characteristics, blood groups, and immunological properties, and even in having the same parasites. For these reasons this family of apes, scientifically known as pongids, are commonly referred to as anthropoid apes. Such a term is misleading in that it suggests that man evolved from a chimpanzee, or in any case from a pongid. Actually the general consensus today is that pongids and hominids are divergent branches of a common ancestral stock that also gave rise to the gibbon. In other words, between 30 and 25 million years ago a common ancestor spawned gibbons, pongids, and hominids, all of which gradually evolved into the modern forms. Each family produced many species. All varieties of hominids became extinct except for the species *Homo sapiens* to which we belong.

The way man's precursors evolved into *Homo sapiens* is difficult to reconstruct, and there are still many gaps in our knowledge. This is because our forerunners lived in forested areas, and conditions there do not favor the preservation of bones in fossil form. The remains of creatures living in more recent times seemed to suggest a sequence of increasingly more advanced humans. In reality many of them were hominids or humans belonging to different species, sometimes contemporaneous, that are now extinct. The hominids can be traced back to 4 million years ago. There is a gap between 5 and 9 million years ago because no remains have been found. But for the period that goes from 9 to 20 million years ago African and Eurasian beds have yielded many fossils of hominoids, that is, members of the superfamily that includes both humans and apes belonging to the common stock.

Ever since he first appeared, man has translated his fast-developing intelligence into works, has had an ever-increasing impact on his environment, and by his actions has made the earth a very different place from the one he found.

For one thing, man has domesticated animals, starting with the wolf, which he made into the dog, his guardian and faithful companion. Later on, after turning from a hunter into a farmer, man domesticated many other animals: cattle, sheep, goats, pigs, chickens, donkeys, horses, and camels, among others. He also took plants that grew in the wild, such as grasses and fruit trees, and cultivated them, developing new strains with greater yields.

The destruction of wild life is a second way in which man has changed his environment. Some animals, the predators, he destroyed because they threatened him or the animals useful to him. Others he hunted for food. For hundreds of thousands of years man did not raise animals but lived on those he managed to kill. Some think that this was the determining factor in the great extinction of 10,000 years ago. Be that as it may, man certainly has caused the extinction of many species during the past 4,000 years, first with a light hand, and then, in the last millennium, in a much more massive way.

A third process that man inaugurated was mass migrations to uninhabited or sparsely populated lands that resulted in the destruction or reduction of the indigenous flora and fauna. These migrations turned what might have remained a limited phenomenon into a worldwide problem. In addition, they altered the ecological balance that prevailed before the arrival of man and the domestic animals he brought with him. Natural shifts in the ecological balance have always occurred, but what we have witnessed in the past few centuries is a man-made change of unusual proportions. In America, for example, cattle, pigs, and horses have supplanted bisons, bears, and cougars, while in Australia the marsupials are giving way to sheep, goats, cattle, and rabbits.

Last but by no means least among the processes of change is man's technological progress, which contains the seeds of its own destruction. In building his machines man is fast using up raw materials that can be recovered only in part when the machines break down. He is depleting the stores of energy that accumulated over millions of years and that cannot be replenished. In brief, by inserting technology into a socioeconomic system that worked for epochs when technology hardly existed, man has initiated an irreversible process of production and consumption that is impoverishing the planet at an accelerating pace.

In the long run, all these processes threaten the very survival of man, a creature who at the beginning was in perfect harmony with the world that had created him but who may soon be at odds with the world he is making.

FROM THE MICROSCOPE TO THE TELESCOPE

Our journey through the earth's past has come to an end. We have observed its birth and its continuous transformations powered by the planet's still hot interior. We have learned that this slow but continuous change in the environment triggers biological evolution, which in turn modifies the environment itself. We have seen how a condensation of particles scattered in space has become a world full of phenomena, beauty, and life, and how this life, after evolving into increasingly more complex forms at an accelerating rate, has culminated in the most intelligent being we know.

It is impressive to see the road traveled on earth in over 4 billion years by part of the material that formed it: from lifeless matter to life; from simple life forms (unicellular beings) to complex life forms (multicellular beings); from a life restricted to water to one encompassing the whole planet; from a sluggish life (cold blood) to an active life (warm blood); from an active life (warm blood) to intelligent life (man). All this occurred in a generally slow but continuous alternation of harmony-disharmony-new harmony with the environment that modified and stimulated life but at the same time was continuously modified by it. What gives more food for thought in this whole series of changes is that there has never been a complete substitution. There still are inanimate matter, unicellular beings, cold-blooded animals, and not very intelligent animals. This makes for a very complex global environment, in which the various forms of activity of matter and life must harmonize ever more, keeping an equilibrium that is continually reestablished as a result of a living world ever richer and more varied. Not only has life progressed from bacteria to humans, but today there is an interaction between bacteria and humans that obviously did not exist when there were only bacteria and single-celled organisms. And the various forms of life interact not only with one another but with the larger entity that harbors them all and is itself alive—the planet earth.

It has been a long journey but it has shown us many things, even though we have merely glimpsed the essential features of our long history. On the other hand, what we have missed in this journey is a small loss compared with another loss that we can appreciate only by considering how much has happened in the earth's past. The point is, we have explored only this one planet, which to us seems quite large and rich but is just a grain of sand in our immense universe. We have lingered on this planet partly because it is our own, but above all because we know it well and we can reconstruct its history for the past 4.5 billion years. But it is as if we had observed under the microscope a tiny fragment of the universe and watched what happened on it in all that time.

Now let us lift our eyes from the microscope and look at the sky. We shall see many stars—very many indeed, yet they are but an infinitesimal part of all the stars that exist, which the telescopes show us as if clouds. Now think of all the clouds of stars scattered in the vastness of space and try to imagine about ten times as many planets circling about most of them. Then you will realize that if even one planet out of ten were alive like our earth, all we have observed under the "microscope" would be multiplied by billions upon billions of cases, perhaps similar in structure and evolution, but among which we would hardly find two alike.

The earth in time: unique in the entire universe and yet an example of what may be happening everywhere on an inconceivable number of worlds.

5 THE END OF THE WORLD

We saw earlier that the stars are born and die. Like every celestial body in the universe, the earth, too, was born. Now it is living, and one day it will die. But how? Centuries ago, when people believed the universe to be very small, the entire world consisted of the earth and the celestial bodies revolving around it—the sun, the moon, the planets, and the glittering stars "fixed" on a crystal sphere. Everything was subordinated to the earth and existed for its benefit, or more precisely, for the benefit of mankind, and by the end of the world one meant the end of the earth and all the celestial bodies.

But the earth is only a speck of matter, even though alive, in an unremarkable corner of space, and the universe would notice its passing less than mankind notices the passing of an ant squashed by a passerby. Today the expression "end of the world" has essentially three meanings. The first is the end of the universe; the second and more restricted in the end of the earth (sometimes including the entire solar system); and the third and yet more limited is the end of the current state of affairs on our planet—and in particular, the end of mankind. The future of the universe and its possible end we shall put off until later. Right now we shall explore the near future up to the time when our sun will no longer be able to warm the solar system. That will not happen until 5 billion years from now. Humankind will certainly end long before then, and the earth itself might not last that long.

THE END OF THE EARTH

A CATASTROPHE

Some time in the near future a group of men are working on a moon base. The earth, motionless in the dark and starry lunar sky, shines like a multicolored disk, prevalently blue. It is always beautiful to see it around the phase of "full earth," with its oceans, its continents, and, here and there, the white smudges of the clouds of ever-changing aspect and extension. One of the men stops working for a moment to lift his head encased in the protective helmet, looks at it, and thinks of his home and the family over there that he will see in a few days when a new team will arrive to relieve them. In that moment of rest and meditation the man's attention is caught by a bright dot moving across the sky toward the earth—too bright to be a spaceship, it can only be a celestial body, a comet or an asteroid passing very close to the earth. In

a short time it will disappear behind the terrestrial disk or perhaps pass
in front of it if its orbit runs between the earth and the moon. But that
is not what happens. The body, already so bright, abruptly increases
in brightness as if it had burst into flames. It is not moving fast now; on
the contrary, it seems almost motionless in front of the earth's disk. By
now it is obvious that the body is falling to the earth. The man looks
on, stilled by disbelief and terror, but it is a question of seconds. Somewhere on the planet there is a blinding flash of light; bright dots flare
up around it and go out almost immediately; then a sort of opaque veil
begins to spread from the impact area; as the hours go by it covers the
whole disk, smudging contours and colors and finally hiding everything
under a dark mantle. By now all the men of the lunar base are outside
watching. A few words are exchanged, but the tragic reality is clear to
all: The earth has been struck by an asteroid.

When they return home, the men find a planet in turmoil and partly
destroyed. The asteroid has fallen on a continent and has formed a
crater 100–200 km in diameter. The atmosphere is full of dust because
the impact has pulverized the asteroid and a piece of the earth's crust.
The sun's rays cannot penetrate this dark cover, and the surface of the
planet is cold and dark. The cold, however, does not last long. In a
short time the dust that blankets the earth begins to warm up. While
part of the energy it receives from the sun is reflected back into space,
part is emitted downward as infrared radiation, that is, heat. The temperature of the atmosphere begins to rise because the same layer of
dust that prevents sunlight from coming through prevents the atmospheric heat from escaping. The temperature will rise until the dust
has settled back on the ground or has dissipated in space. And if this
does not happen in a reasonably short time, the air will get hot enough
to kill all the beings that had survived the impact and its disastrous
consequences, such as earthquakes and hurricanes. (If the asteroid had
fallen in the ocean, there would have been tidal waves and widespread
flooding.) G. C. Reid has calculated that if the dust layer formed by the
collision with an asteroid 10 km across were to obscure the sun for a
few years, the temperature of our atmosphere would rise hundreds of
degrees. In sum, after such a catastrophe the earth would look very
much like the planet Venus and life might be completely obliterated.

Now we are wondering whether this could actually happen, and the
answer unfortunately is affirmative. It could happen tomorrow or in a

few minutes, all of a sudden. It could be just as sudden as the fall of the comet that struck Siberia in the early morning of 30 June 1908. In fact, there are bodies in the solar system that can cause catastrophes such as those we have just described.

THE EARTH-GRAZING ASTEROIDS (EGAs)

Among the thousands of asteroids that orbit the sun between Mars and Jupiter, some travel different paths from the majority (figure 61). Much like comets, these "individualists" often (but not always) have very elongated elliptical orbits that periodically thrust them toward the sun and that may penetrate the earth's orbit or have their perihelia just outside of it. In the first case, if the asteroid's orbit were to lie on the same plane as the earth's orbit, it would cross the latter at two points. Normally this does not happen because the two orbital planes do not coincide; hence the two points in the earth's orbit that would correspond to intersections lie a little above or a little below the orbit of the asteroid. However, the inclinations of the orbital planes are generally very small, and consequently these asteroids often pass quite close to the earth.

The first of the group, discovered 24 April 1932, was named Apollo. Computations showed that the asteroid's orbit had the strange characteristic of running inside the orbit of Venus. After the discovery, we lost sight of Apollo until 1973; then in 1980 it passed 7.5 million km from the earth. That may sound like a considerable distance, but by planetary standards it is not. Remember that our average distance from the sun is 149.6 million km and that the nearest planet, Venus, never comes closer than some 40 million km. Furthermore, another such asteroid, Adonis, was discovered 12 February 1936. Computation of the orbit showed that 5 days earlier it has passed just 2 million km from the earth. Adonis also was lost, and it was not seen again until 1977. On 28 October 1937, astronomers found the third of these asteroids, Hermes, which 2 days later passed only 800,000 km from the earth, a little more than twice the moon's distance from us. That was too close for comfort, especially when calculations showed that the minimum distance between the earth's orbit and Hermes's was even less than 800,000 km. If one day the two bodies should pass at the same time through the two points of minimum distance, they would be only 300,000 km from each other, which is less than the distance separating us from the moon.

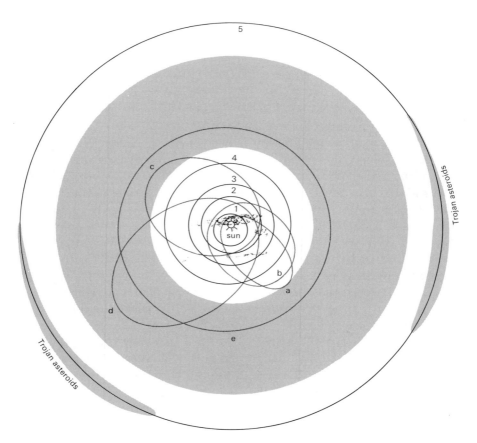

Figure 61   Orbits of some asteroids and planets projected on the plane of the
earth's orbit: (a) Icarus, (b) Hermes, (c) Apollo, (d) Adonis, (e) Ceres; (1) Mercury, (2) Venus, (3) the earth, (4) Mars, (5) Jupiter. Compare the orbit of Ceres,
which is a normal asteroid, with the orbits of the EGAs. [From *Scientific American* (April 1965)]

These bodies, which normally are tens of millions of kilometers away from the earth, but occasionally pass close to us, are called Earth-Grazing Asteroids, or EGAs for short. At present, about 40 of them are known, the most distant being Ganymede, which can approach within 51 million km of the earth.

It may seem exaggerated to worry about bodies that generally do not come closer than some millions of kilometers, even granting that such distances are fairly small on a cosmic scale. We do not feel in any great danger when we cross another car with barely a meter to spare. The trouble is that the inclinations of the orbital planes are not stable, and as we shall soon see, the distance between the orbit of an asteroid and that of the earth may even fall to zero. Furthermore, in the case of asteroids whose orbits lie entirely outside the earth's, we know that when their perihelia are very close, the orbits can quickly change so as to cross our own. Thus some day an EGA might travel an orbit exactly intersecting that of the earth, and if both bodies were to pass through the intersection at the same time, a collision would be inevitable. In fact, not too long ago we barely escaped such a collision.

On the afternoon of 10 August 1972, a large number of astonished people saw a bright bolide that crossed the North American sky from Utah to Alberta, Canada. It was a very bright meteor that many people photographed and filmed (figure 62). When it passed over Montana the sky rumbled.

Among the thousands of people who saw the bolide was L. Jacchia, an astronomer of great versatility and with a particular expertise in the field. After enjoying the spectacle like everybody else, he began to collect information and was particularly struck by two facts: first, the thunder that shook Montana was not heard in Alberta, where the bolide disappeared, and, second, there was no sign of it falling anywhere, whereas the impact of such a bright body should have released the energy of an atomic bomb. Due to the decrease in atmospheric temperature above 60 km, the physics of the atmosphere dictates that if the bolide had flown above that level, its shock waves would have been refracted upward; hence no thunder would have been heard. Jacchia concluded that the asteroid had literally grazed the earth, penetrating its atmosphere to a minimum height of less than 60 km and causing the thunder heard over Montana; then it had continued to describe its orbit around the sun without striking our planet.

Figure 62   Photographs of the asteroid that grazed the earth on 10 August, 1972. It did not hit our planet, but came close enough to penetrate its atmosphere. The bolide was seen by thousands of people. According to calculations, it was 80 meters across and weighed about a million tons. [Courtesy Mr. and Mrs. James Baker, Lillian, AL]

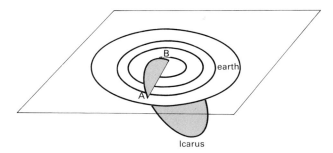

Figure 63   The orbit of Icarus, one of the EGAs. The line of intersection (A–B) of the orbital planes of Icarus and the earth rotates with the passage of time. Hence the orbit of the asteroid periodically crosses that of the earth. If both bodies were to pass through that point at the same time, a collision would occur. Thus the event is possible, though extremely unlikely.

When its orbit was computed from a large number of observations, the bolide turned out to be an EGA that had crossed the earth's atmosphere at an average speed of 15 km/sec. Assuming that at a height of 100 km its magnitude was − 19 (about 400 times brighter than the full moon), it would have had a diameter of 80 m and a mass of 1 million tons, 25 times the mass of the small comet that crashed in Siberia in 1908.

The passage of this body, very small but already fairly dangerous, demonstrates that there certainly are EGAs of whose existence we are ignorant that could strike quite suddenly. E. M. Shoemaker has estimated that there are from 750 to 1,000 EGAs with a diameter greater than 1 km. Hence the great majority of them are unknown, and for all we know many of them might have orbits that cross the earth's path. Among the known ones, none is currently on a collision course with the earth. Because of the attraction of the major planets, however, the orbital plane of each asteroid slowly rotates. Consequently, at regular intervals of a few thousand years, each orbit crosses our own (figure 63). Fortunately, when the asteroid passes through the intersection the earth is usually in another point of its orbit. The likelihood that the earth and the asteroid will arrive at the intersection at the same time is very small, only once in 200 million years. If Shoemaker's estimate of the number of EGAs is correct, our planet should be struck by one of them on the average every 250,000 years.

As the existing asteroids strike the earth or other planets, one would expect that their number should gradually decrease, thereby reducing the chance of a collision. But studies of the lunar craters show that in the past 3 billion years the flow of projectiles capable of forming craters has remained nearly constant. Evidently there are new EGAs, some coming from the vast reservoir of asteroids in orbit between Mars and Jupiter due to a change of orbit, others being the surviving solid nuclei of extinct comets.

If it is true that once in a while an EGA strikes the earth, we should find evidence of past impacts—not for all of them, of course, because 70% of the bodies are bound to fall in the oceans. These may cause widespread flooding, but after that terrible yet brief disaster is over, no sign of impact would be left. The bodies that fall on the continents leave scars that last far longer, though in time they are erased by atmospheric agents. Craters more than 20 km across can be seen as much as 600 million years after the impact. R. A. F. Grieve and P. B. Robertson have compiled a list of 78 geologic features probably due to hits by EGAs. Three of them are very ancient: the oldest dates to 2 billion years ago; the others, to less than 600 million years ago. These craters range up to 140 km in diameter, but the original ones must have been much larger and deeper, because the edges have been eroded and what is left now is only the central, deeper part of the crater.

There is no doubt that the fall of an asteroid can cause immense damage—total destruction for a radius of hundreds of kilometers, intense seismic activity, and tidal waves capable of engulfing an entire fleet. It would also be followed by clouds of dust dense enough to obscure the sun, resulting in large variations in temperature and great loss of animal and plant life. Every impact was and will surely be a terrible catastrophe, but ordinarily a local and short-lived event, leaving no permanent traces for geologists or paleobiologists other than a crater.

From the evidence of past hits we can conclude that collision with an EGA would be a disaster but would not cause the end of the world. But one could think of larger EGAs, perhaps as big as the moon—so large and massive that a collision could break our planet apart. But there is no need to worry about this eventuality. The number of unknown EGAs is so high because those that have eluded us so far are almost exclusively the smallest ones. If you recall, nearly all of the largest ones fell before 3.7 billion years ago. They were the last planetesimals; the remaining ones or their fragments are only "crumbs."

Thus we can be fairly sure that the earth will not end in this manner. Discounting the extremely unlikely event of a collision with an object from outside the solar system, our planet will never be struck by a body large enough to pulverize it. Once in a while it will be wounded by minor bodies; it might be horribly devastated by one of the larger EGAs if its path should cross our own; but it will not be destroyed, and its wounds will gradually heal as they have done in the past.

OTHER CATASTROPHES

One can think of less "personal" catastrophes that could thoroughly destroy not only the earth but other planets as well. Up to a few decades ago astronomers still entertained the possibility that the sun might explode as a nova or a supernova. Today we know that neither event will occur. As we have seen, stars the mass of the sun never become supernovae because their evolution follows an entirely different route. As for the nova explosion, it is a phenomenon peculiar to binary systems, and the sun, needless to say, is a single star.

Another possibility would be the encounter of the sun with another star. Even if it stopped short of an actual collision, a close encounter would have disastrous consequences. Apart from increases in temperature, the tidal effects on the sun could be large enough to cause the ejection of fiery matter that upon reaching the inner planets would undoubtedly destroy them. Although possible, this event is exceedingly unlikely. If you recall, taking into account the number of stars in the Galaxy and the vastness of space, Struve calculated that the chance of an encounter is very low. Recently, Tai Wen-sai redid the calculations on the basis of the latest data and found a lower value for this probability. According to the Chinese astronomer the encounter of the sun with another star could occur once in 2.7 million billion years, which is a span of time more than 100,000 billion times the age of the universe. This does not mean that it could not happen a thousand years from now; it just means that it is highly unlikely ever to occur. In any case, it will not happen in the next few years because there is no star in our neighborhood that is set on a collision course with the solar system.

THE END OF MANKIND

Even though the earth itself will not be destroyed, there could be a ca-
lamity capable of obliterating life on the planet—possibly not all of it,
perhaps just the human race. One thing is certain: Mankind as we
know it today will end before the earth. As we have just seen, there is
hardly any chance that the earth will break apart from the collision
with a large body. Barring that, there is no reason why the earth should
not continue to live its planetary life of continuous change and evolution. In that sense it could last until the sun is radically different some 5
billion years from now, which is longer than the earth has lived thus
far. Considering how much terrestrial life has changed just in the past
800 million years and how much the pace of evolution has accelerated,
it is easy to predict that the humankind we know today cannot possibly
last as long as the earth. The question is a different one: Will mankind
die out and pass on mastery of the planet to one species after the other,
or will mankind survive and continue to change, perhaps extending its
ecological niche into space, either on habitable worlds or in man-made
space colonies? Even if man should survive, judging from past biological evolution it seems fairly certain that tens of millions of years from
now our descendants will be as different from us as we are from our
mammalian ancestors of the dinosaur age.

First of all we shall look at some future scenarios from which we may
be able to deduce if and when mankind, or even all of terrestrial life,
might end. We shall not consider phenomena such as glaciations, reversals of the magnetic field, or continental drift, which have occurred
repeatedly in the past and will continue to occur with similar consequences. Rather, we shall look at new facts and at some events that may
not be new but might have a different outcome in the future.

CHANGES ON THE SUN

It is generally assumed by scientists that the sun has been the same
throughout geologic time. This conviction is well founded, since for
over 4 billion years the sun has been on the main sequence, and in this
phase stars do not change, or change very little. It is true that there are
variations in solar activity, which, among other things, cause its appearance to change from one day to the next. But this is a cyclical phenomenon that has a maximum and a minimum every 11 years. Thus, even

though the sun's appearance varies, this happens with a certain regularity. On a human scale, the sun is much like a man who does different things every day but always expends the same amount of energy and alternates periods of activity with periods of rest. But there is one difference—a man replenishes his lost energy by eating; the sun, instead, has to dip into its own mass to produce energy and, like anyone who fasts, loses a certain amount of mass every day. This has been the general belief ever since the sun's 11-year cycle was discovered around the middle of the last century.

In 1893, however, a solar astronomer at Greenwich Observatory, E. W. Maunder, made an extraordinary discovery while consulting some old publications. According to the observations made since Galileo's time, there had been a period of 70 years, from 1645 to 1715, during which the sun had appeared practically spotless. For periods as long as 10 years not a single sunspot was seen, and over longer intervals (30–60 years) observers recorded no more than a group of spots at a time (figure 64). In brief, Maunder found that the total number of sunspots in the 1645–1715 period was smaller than the number of sunspots that were then being seen in one year of average activity. Maunder published his findings in 1894 and again in 1922, but the astronomical community was not impressed. No doubt, this interruption in the solar cycle was bothersome for the theoretical interpretation of solar activity, and it was more convenient to regard Maunder's discovery as the fancy of an honest but overenthusiastic researcher who had placed too much credence in old observations.

In 1976 Maunder's cause was taken up by the solar physicist J. A. Eddy. Fifty years of advances in solar astronomy had shown that sunspots are not the only manifestation of solar activity. Consequently, besides rechecking the seventeenth-century observations of sunspots, Eddy also looked into other phenomena. He not only confirmed the solar minimum discovered by Maunder, but found other minima and moreover charted the variations in solar activity over a few millennia.

The first thing Eddy did through his extensive historical research was to prove that the absence of sunspots was not due to a lack of observations; in fact, the seventeenth-century astronomers themselves had considered it very strange. Eddy also sought additional evidence from the naked-eye observations made by the Chinese, Japanese, and Koreans. Unlike the Westerners, they were not conditioned by the

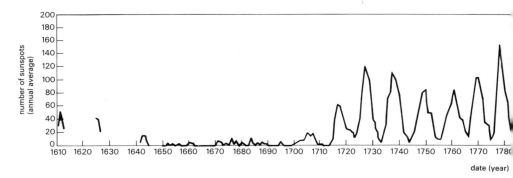

Figure 64   The 11-year cycle of sunspots from 1610 to 1976. The first three segments have been reconstructed from the observations of Galileo, C. Scheiner, and J. Hevelius. Note, beside Maunder's minimum, the different heights of the peaks during the period regarded as normal. [From *Le Scienze* (September 1977)]

myth of the sun's flawlessness, and in their chronicles they recorded the dates of appearance of sunspots large enough to be seen without telescope. From 28 BC to 1743 AD the Easterners noted about 10 spots per century but none at all from 1639 to 1720, that is, a period of time encompassing Maunder's minimum.

Another indicator of solar activity is the splendid corona that can be seen with the naked eye during total eclipses of the sun. At times of maximum solar activity the corona appears roughly circular in form, with many plumelike rays radiating outward like petals of a dahlia. Around the solar minimum the polar rays are shorter, while there are long streamers in the equatorial regions. Not only the corona's shape but its very existence depend on solar activity. If that ceased, after a few years only a false corona would be left that during a total eclipse would appear as a faint reddish halo around the dark disk of the moon covering the sun. Between 1645 and 1715 there were 63 total eclipses of the sun, many of which were observed mainly in Europe. Checking the descriptions of the sun during these eclipses, Eddy found no mention of the silvery halo familiar to us, but only of a faint and thin reddish ring.

Solar activity also manifests itself on the earth, notably with the spectacular auroras. Between 1645 and 1715 very few Northern Lights

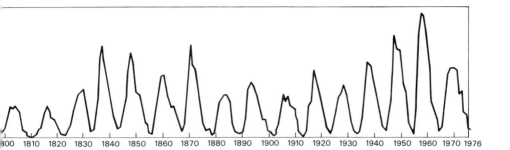

were observed even in Scandinavia, where at present they are seen almost every night, and for a 37-year period no aurora was seen anywhere on earth. At long last one was seen in England in March 1716. In his sixties, the famous astronomer Halley declared that he had never been able to see one and wrote an article on the subject. The disappearance of northern lights was another confirmation of Maunder's minimum.

Eddy also searched for evidence of solar activity not in man's memory or chronicles but in something more objective, lasting, and extended in time. He found it in the rings that mark the annual growth of trees. High in the atmosphere, cosmic rays produce an isotope of carbon (carbon 14) that subsequently descends to soil level and accumulates in the outer layer of wood grown by the trees every year. The flux of cosmic rays is modulated by solar activity, which modifies the extended magnetic field of the sun. During the periods of maximum solar activity the flux diminishes and very little carbon 14 is produced. Thus, by analyzing the amount of carbon 14 in each ring of an ancient tree it is possible to chart solar activity for hundreds of years. Eddy was able to chart it for the past 5,000 years (figure 65). He found not only Maunder's minimum and a previous one from 1400 to 1510 (which he called Spörer's minimum), but also several other minima occurring around 700 AD (medieval minimum), 400 BC (Greek minimum), 700 BC (Homeric minimum), and 1300 BC (Egyptian minimum). These minima seem to coincide with as many small glaciations. From the same set of data Eddy also discovered that there have been times of exceptionally high solar activity, namely, around 1200 (medieval maximum), at the beginning of the Christian era (Roman maximum), around 1800 BC

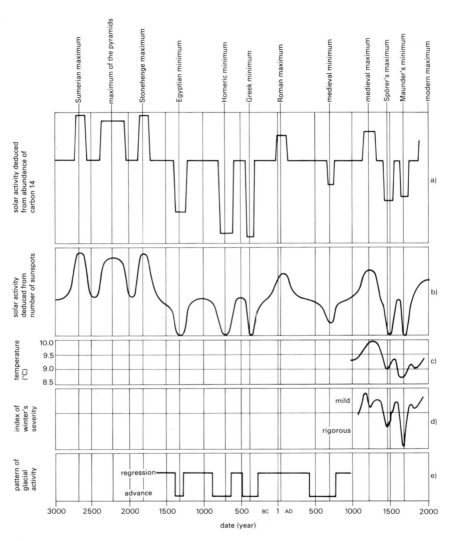

Figure 65   Solar activity since 3000 B.C. as deduced from the abundance of carbon 14 in the growth rings of *Pinus Aristata:* (a) the observed abundance, calibrated in terms of solar variations on the basis of Maunder's minimum; (b) variation in solar activity deduced by Eddy from the above; (c) an estimate of the average annual temperature in England; (d) rigor of the winters in Paris and London; (e) times of retreat or advance of the alpine glaciers. [From *Le Scienze* (September 1977)]

(Stonehenge maximum), 2200 BC (maximum of the pyramids), and 2700 BC (Sumeric maximum). Currently we seem to be on our way to a new great maximum that should reach its peak around 2100 or 2200.

Eddy's findings are significant in that they show that solar activity has far deeper roots than suspected from the 11-year cycle and, more important, that solar activity varies over much longer periods of time and with greater amplitude, perhaps affecting the earth's climate. After this discovery one cannot exclude the possibility that larger fluctuations may have occurred in the past or may occur in the future. If such oscillations were to be greatly amplified, the consequences for our living world could be very serious indeed.

THE SUPERNOVAE

The cosmic explosion that occurred about 65 million years ago 880 light-years from our planet was almost certainly the cause of the ecological disaster that at the same time marked the end of the dinosaurs and the beginning of the age of mammals. Its effects in space, the only aspect of the event observable today, lead us to believe that it involved either a body of exceedingly large mass or a chain of supernovae. Such mighty explosions, also called super-supernovae, are believed to be very rare. Hence it is extremely unlikely that one will take place right near the earth in the conceivable future.

Nevertheless, our planet could be engulfed by a lethal wave coming from a normal supernova relatively nearby. A few years ago W. H. Tucker calculated that a supernova 60 light-years away with an intense emission of x- and γ-rays would immediately damage the protective layer of ozone in our atmosphere and would have lethal effects in the days and years to come through other phenomena of interaction with the radiation and material it ejected. Recent x-ray observations from scientific satellites have not borne out Tucker's predictions of an intense x- and γ-ray emission from supernova explosions. However, Tucker's model assumes that most of the x- and γ-radiation, with an energy of 3×10^{47} ergs, would leave the supernova within 3 hours of the explosion. During this period the expanding supernova shell is still opaque to light. Thus, when the x-ray detector is aimed at a supernova discovered optically, the bulk of the x- and γ-radiation has already gone by.

In 1979 R. T. Rood and his collaborators made an interesting discovery while studying the concentration of the ion NO_3^- in the ice of Antarctica. This ion is produced in the earth's atmosphere by the bombardment of x- or γ-rays. The ions that form over the centuries, in amounts proportional to the intensity of the radiation, have accumulated in layers on the southern icecap. Thus, by measuring the concentration of NO_3^- ions at various depths it is possible to estimate the intensity of the x- or γ-ray bombardment in the past. Rood and his coworkers had cored the ice to measure the variation in the x- and γ-ray emission from the sun. But in addition to more or less wide fluctuations, they found three intense peaks of very short duration corresponding to the years 1604, 1572, and 1181. In those years three supernovae had appeared. Unfortunately, the coring was not deep enough to reach levels corresponding to the very intense supernovae of 1054 (which formed the Crab nebula) and 1006. In any case, Rood and his collaborators think that the explanation for the observed peaks is that each of the three supernovae emitted about 10^{50} ergs, in the form of x- or γ-rays. Such an emission is 1,000 times higher than that predicted by Tucker, which would be lethal to terrestrial organisms if it originated from a supernova 60 light-years away.

If Rood's results were confirmed by other findings—for example, by the discovery of high radiation levels at the time of the 1054 or 1006 supernovae—predictions of an intense x- and γ-ray emission from supernovae would also be confirmed. Should it turn out to be 1,000 times more intense than Tucker predicted, it would be lethal even if it came from a supernova 10 times farther away than 60 light-years. In other words, a supernova explosion would have lethal effects within a much larger radius, and the chance of its affecting the earth would be that much greater.[37]

SELF-DESTRUCTION

All we have seen thus far, from the evolution of the stars to the birth of the solar system, is based on observations of events that occur at present elsewhere in the cosmos, while the history of the earth is a plausible reconstruction of the past on the basis of geologic evidence. On the other hand, events that have never happened before are very difficult to predict for the very reason that they are unprecedented. But there is one event that is less difficult to foresee because it may not be too far off and because we would be both its victims and its agents—and that is the end of mankind through self-destruction.

We are living at a unique and decisive time in the history of the human race. I know this has been said before, but now there is a difference. Our history has seen many turning points, but even the most dramatic, like the fall of the Roman Empire or the destruction of the ancient American civilizations, did not involve the entire planet. Today it is different: What could happen is simply that mankind might no longer have a history.

We are all aware that the human race could be wiped out by a nuclear war that could start at any moment, perhaps by mistake (the weapons are already armed and aimed). Everybody hopes this will never happen, and possibly it never will. We also are familiar with the warnings of an impending crisis due to the exhaustion of raw materials and energy sources, something that at the current rate of consumption is bound to happen in the near future. So many books and articles have been written about these issues that it seems useless to discuss them again in these few pages. But unfortunately, there are other threats to our survival. Dangers also exist as a consequence of our technological progress and the very prosperity we enjoy, more subtle dangers, but equally evident if one looks around carefully.

Take a look at figure 66, which shows the number of people in the world, the speed of transportation, and the speed of communications in the past 9,000 years. Similar curves would be obtained for everything else that has to do with man, from the speed of performing mathematical calculations to the speed with which new ideologies are born and spread, from the number of important discoveries made every year to the amount of leisure time available on the average to every human being. What strikes you immediately is that the general rate of increase is very low for thousands of years and then, from about 1800, accelerates rapidly, reaching today's frenetic pace. This rate of growth is what we call progress, but at the same time it contains the seed of the monster that could devour everything.

I shall give you a practical example. At the time of the Roman Empire a letter normally took 3 days to go from Rome to Naples and 15 days to go from Damascus to Rome. Around the turn of the century a person who lived in Perugia wrote to a friend in Naples to arrange a

RELATIVE SIZE OF THE WORLD BASED ON THE REDUCTION IN TRANSPORTATION TIME

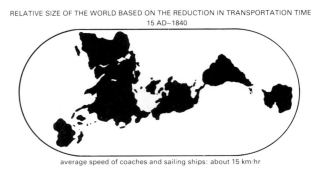

15 AD–1840

average speed of coaches and sailing ships: about 15 km/hr

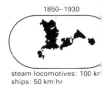

1850–1930

steam locomotives: 100 kr
ships: 50 km/hr

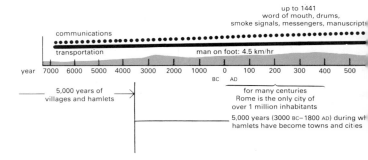

Figure 66   The increase in world population (shaded area), speed of transportation (solid/dotted line), and speed of communications (dotted line) from 7000 BC to the present illustrates the uniqueness of the modern era in the history of mankind. Dramatically evident is the steep upward turn that all the curves show starting from the beginning of the last century. [From Center for Integrative Studies, *World Facts and Trends*, John McHale, 1977]

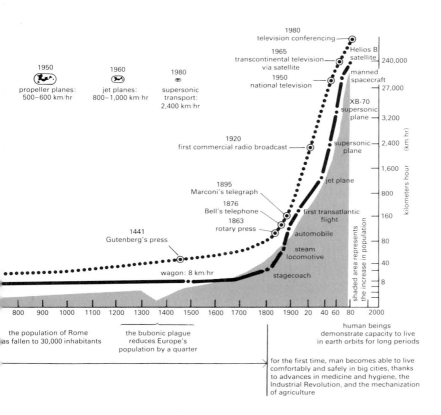

1980
television conferencing

1965
transcontinental television
via satellite

Helios B
satellite — 240,000

1950
national television

manned
spacecraft — 27,000

XB-70
supersonic
plane — 3,200

1950
propeller planes:
500–600 km/hr

1960
jet planes:
800–1,000 km/hr

1980
supersonic
transport:
2,400 km/hr

1920
first commercial radio broadcast

supersonic — 2,400
plane

1,600

800

1895
Marconi's telegraph

jet plane

1876
Bell's telephone

160

1863
rotary press

first transatlantic
flight

1441
Gutenberg's press

automobile

80

steam
locomotive

40

wagon: 8 km/hr

stagecoach

8

kilometers hour   (km hr)

shaded area represents
the increase in population

800 900 1000 1100 1200 1300 1400 1500 1600 1700 1800 1900 20 40 60 80 2000

the population of Rome
as fallen to 30,000 inhabitants

the bubonic plague
reduces Europe's
population by a quarter

human beings
demonstrate capacity to live
in earth orbits for long periods

for the first time, man becomes able to live
comfortably and safely in big cities, thanks
to advances in medicine and hygiene, the
Industrial Revolution, and the mechanization
of agriculture

meeting in Rome 2 days hence.[38] Thus a letter was expected to travel from Perugia to Naples or vice versa in less than 2 days. This can still happen today, but it is an exception. On the average our mail takes longer now than it did at the turn of the century, and on certain routes today's letters take even longer than they did 2,000 years ago.

Why? The answer is also contained in figure 66. The point is that everything has been growing at an accelerating pace. If the speed of communications (dotted line) has increased, so has the number of people who send messages and the difficulty of routing, transporting, and delivering them. Yes, the telephone call can replace the letter, but the telephone system is already overloaded. Millions of cars and trucks speed by on our highways, but a break in the road or a small accident is enough to cause massive traffic jams. Existing structures, be they the postal service or the transportation network, cannot keep pace with the ever-increasing volume of traffic.

If population and the volume of traffic in every sector continue to increase, the situation could deteriorate further for another reason. As R. Vacca pointed out in the 1970s, our technological civilization is heavily dependent on large structures, that is, organizations in which a collective goal is achieved by utilizing a large number of people and machines. Telephones and telegraphs, railroads, airlines, postal services, energy production and distribution, automated industrial processes, and military establishments are only some of the large structures on which we depend. Such a system works only if each of its parts works; like the chain that pulls the cart, if even a single link breaks down, the cart comes to a stop. According to Vacca, if these giants with clay feet should collapse, a new Dark Ages would surely come upon us. And one of the reasons why our structures could break down is their own unlimited growth.

But even the indiscriminate growth of a single factor could have serious, perhaps deadly, consequences.[39] Among the various disasters threatening us I shall mention one that involves our atmosphere, the good old atmosphere that was born so long ago and made it possible for us and our environment to exist.

One of the consequences of our activities and improved standard of living is an increase in the production of heat and gases. This results in an increase in the temperature of the atmosphere for two reasons. First is the direct production of heat—for example, by burning fuels for home heating. Second, less obvious but more efficient and dangerous, is the so-called "greenhouse effect." We have seen this process at work when we discussed the heating of the atmosphere due to a layer of dust blanketing the earth. Carbon dioxide and water vapor, two by-products of fuel combustion, behave more or less like dust in that they cause the atmospheric temperature to rise. Obviously, this effect becomes more pronounced as the amounts of carbon dioxide and water vapor increase. Levels have been rising rapidly since about mid-nineteenth century due to the burning of coal, oil, and natural gas, which, besides heat, produce carbon dioxide and water vapor. Of all the carbon dioxide spewed out by chimneys, smokestacks, and the exhaust pipes of cars, trucks, and planes, half is absorbed by oceans and forests, but the other half is trapped in the atmosphere, raising its temperature.

Figure 67 illustrates the increase in carbon dioxide in the earth's atmosphere from 1860 to 1975, and predicted increases to the year 2035. The graph shows a well-documented upward trend, which is expected to become steeper in the future. There are four curves because different authors have obtained different results, but they all agree that a steep rise is inevitable.

From the increase in the amount of carbon dioxide in the atmosphere one can estimate the rise in surface temperature due to the greenhouse effect and compare it with the mean global temperatures recorded since 1860 (figure 68). As I mentioned earlier, the earth is coming out of a mild interglacial period, and for the past few thousand years there has been a downward trend in the earth's average temperature. Now it appears that this cold trend will have no effect on the climate of the next century because it will be more than offset by the man-made warming effect, which is 40 times greater than the natural rate of variation. If this warming process does not stop or slow down, the air will become unbearably hot in a relatively short time, perhaps in no more than 200 years. This period of time encompasses at most ten generations, much too few for the human species to adapt to the new conditions through natural selection. And when the mechanism of natural selection does not have the time necessary to produce evolution we have a catastrophe, that is, the extinction of the species that have not had time to adapt to the new environment.

If this catastrophe should occur, the disappearance of man will put an end to the injection of new carbon dioxide into the atmosphere and

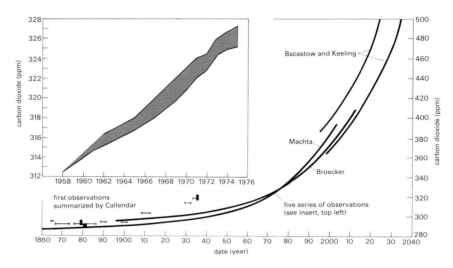

Figure 67   Increase in the amount of carbon dioxide in the earth's atmosphere from 1860 to 1975 and theoretical predictions up to the year 2035. In particular, the diagram shows the results of observations made from four different locations (Cape Borrow, Mauna Loa, American Samoa, and South Pole) and in the course of flights organized by Swedish scientists. The theoretical curves were calculated on the assumption that the burning of fossil fuels will continue to increase at the current, almost exponential, rate; the curves are different because the various authors differed in the ways they took into account the absorption of man-made CO_2 by oceans and biomasses (woods, etc.). The amount of CO_2 in parts per million (ppm) refers to the unit of volume. [From W. W. Kellog, *WMD Bulletin* (October 1977)]

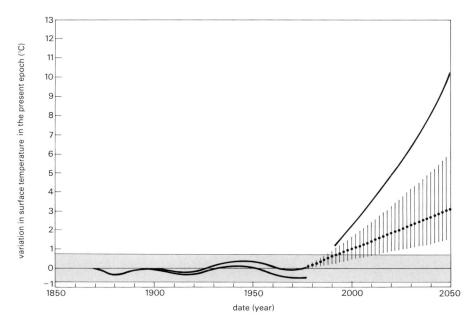

Figure 68   The average surface temperature recorded in the Northern Hemisphere from 1860 to 1980 (solid curve) and what it would have been without
the carbon dioxide produced by man's activities (dashed curve below solid
curve). The two curves calculated for the future correspond, respectively, to
the average temperature (dots) and to that of the polar regions (solid curve
rising sharply at top right). The vertical bars indicate the degree of uncertainty
and define the extremes within which the actual curve could be more or less
steep than predicted. The horizontal bar at the base of the graph defines the
maximum interval of temperature fluctuations during the past millennium, before man began to make large use of fossil coal for home heating and industry.
[From W. W. Kellog, *WMD Bulletin* (October 1977)]

the heating process will stop. The polar icecaps, melted by the heat, will form again; the seas flooding the densely populated coastal plains will subside; the climate will stabilize, and in the space of a few centuries the environment will become the same as it was before. In the life of the earth only a brief moment has gone by, but in that moment the self-proclaimed "masters of creation" have disappeared, and with them, many of the creatures that inhabit today's world.

We have only caught glimpses of our possible future, but I think it is enough for us to conclude that the human race has started processes that are running out of control and will soon reach the critical stage. If we do not take action in time—within the next twenty years, the experts tell us—whether or not such processes ever reach the point of destroying the species, we shall surely plunge into a Dark Ages much worse than the past one. It will then become evident to everyone that maintaining the ecological balance is much more important to the survival of the species than a high standard of living. There is a saying in Italy that poor people always use to console themselves—"Health is all that counts."—which is nothing but the equilibrium of the body.

As we approach these crucial times for mankind and perhaps for all of life, one cannot help looking back over the course of evolution. At this point, while we are about to reach a decisive turning point for mankind and perhaps for all of terrestrial life, the evolution of life since its appearance on earth panoramically flows again in front of our eyes in a flash. Born as a transformation of inorganic matter, life then evolved into forms ever more complex and richer in potential, eventually to develop into a mind and a spirit that look back to discover the road traveled and look forward to try to read the future; a mind and a spirit capable of discovering and creating by means of the intuition, reasoning, and pure abstraction of the mathematician, the philosopher, and the artist; a mind and a spirit whose discoveries, creations, and thoughts are a source of intoxicating pride and deep anguish—anguish because after the exhilaration of discovery or the creative moment one realizes that all that one has accomplished always falls short of expectations, that no matter how much one learns, one can never know everything, and therefore one can never explain the universe.

But by now man is sure of at least one thing: Man is not a fallen angel to be commiserated, but one of the results of a universal evolution that has been going on for billions of years. Man knows that he himself can make a contribution, though minute, to the general evolution toward an unknown goal. And he also knows that if he does not do it, if he cannot be a part of the general evolution of nature, he will be discarded, because this ascent is not stopped even by catastrophic events—as demonstrated by the entire past of the earth. And this responsibility, that he is Adam confronting daily the Tree of Good and Evil, exhilarates and torments him.

THE INHERITORS

If mankind cannot regulate itself and make the quantitative jump required by the present circumstances—if it should succumb because of self-destruction, impoverishment of the planet, or population explosion—it does not necessarily follow that all life will vanish from the face of the earth. The departure of the human race might leave behind a squalid and desolate planet, a planet perhaps sparsely populated but not entirely devoid of life. Then, among the surviving species, almost certainly at least one will develop that, being best suited to the new environment, will emerge over the others.

Who will be the new masters of the world? It certainly will not be animals out of the past, like amphibians or dinosaurs, which played their part in evolution and then for one reason or another declined, died out or changed. Nor will it be any of the species whose lives are closely linked with man; these will be dragged along in his fall. And as for wildlife, which man has already ravaged, I fear it will fall victim to population explosion, that is, man's fight against hunger.

Strange as it may seem, an examination of the past and the present shows that our most likely inheritors will be the insects. Indeed, according to most entomologists we already are in the age of insects. This startling idea sounds more plausible when you realize that 900,000 species of insects are currently known to inhabit the earth. It is an enormous number, especially if you consider that it is larger than the number of all other animal and vegetable species combined. The number of individual insects is also prodigious; it is estimated that there is over a billion billion of them. Considering that in 1980 the world's human population numbered 4.5 billion individuals, that makes 220,000 insects for each human being on earth. Furthermore, ever since the insects first appeared about 350 million years ago, both the number of species and the number of individuals have been continuously growing.

Though less impressive than sheer numbers, this latter fact is more worrisome.

First of all, with so many species of insects in existence there is a good chance that one of them will be able to adapt to the new environmental conditions. In the second place, if insect evolution has not faltered in more than 300 million years, it means that either the environment has always been favorable to them or that they have a remarkable power of adaptation. The case is most likely the latter, because the environment has changed several times in the past and has often become quite hostile to them. Just think of the latest ordeal they have had to endure, namely, the all-out war that man has waged against them with insecticides. Despite the veritable massacre of some species, the class as a whole is emerging unscathed, partly because natural selection, accelerated by the insecticides, has quickly produced resistant strains, and partly because of the development of species previously kept in check by natural enemies that have been decimated by the insecticides.

Apart from this, insects are better qualified to inherit the earth than most other living beings. Their muscular strength is hundreds of times greater than man's; they are largely covered by protective shields; they are able to broadcast and receive electromagnetic waves and ultrasounds; and they have eyes capable of seeing in many directions at once, sensitive to the ultraviolet, and efficient even in the dark of night or in underground tunnels. Of course, man has many of these abilities, too, but he has acquired them by using his intelligence to build instruments and machines. Also, the insects have a highly developed sense of smell, which in man is little more than rudimentary and which he has not cared to improve by means of machines. Given such natural resources, there is no telling how far some species of insects could go if they acquired a more powerful brain.

And that is not all. Some species, such as bees, ants, and termites, have a better social organization than man's. Also, like most other species on the planet, they are more in tune with nature than we are. As soon as it was in his power, man has harnessed, fought, or destroyed the rest of the living beings in the animal and vegetal kingdoms, whereas the insects have always lived as one among equals, which benefits both them and the environment. Naturally, they do not do it for a precise purpose or as a matter of conscious policy. But they do it. In the past, for example, the leaf-eating insects stimulated physical and chemical defense mechanisms in the plants that were very beneficial to the plants themselves. Pollinating insects like butterflies, bees, and many others have been a factor in the evolution of flowering plants. To such insects we owe the shapes, colors, and scents of flowers—in other words, the beauty that plants developed in order to make them attractive millions of years before the woman was born to whom a man would offer a rose.

Clearly this prediction of a future kingdom of the insects is only one of the scenarios that can be envisaged. Perhaps man will not be replaced by the insects. If he survives the coming crisis and grows up, perhaps it will be the human race that will continue to evolve, or perhaps new species generated by man in some way—for example, through genetic engineering.

So much fantasy, so much science fiction can be developed around this theme! We, too, have not been able to resist the temptation. But even keeping within the bounds of reasonableness, one can still make two general predictions. First, judging from the past, tens to hundreds of millions of years from now the species that in turn will rise to world supremacy will be as different from the present human species as the dinosaurs were from primitive fish and the mammals from the dinosaurs. And second, if what occurred in the past continues to occur, those species will be increasingly more complex and versatile—in other words, much more advanced than our own.

THE END OF THE SOLAR SYSTEM

Humankind will perish or evolve. Our successors and the successors of our successors will rise and fall. New continents and oceans will form. The mountains that today rise in all their majesty will be worn down by rain, ice, and wind, and new ranges will be created by the clash of new continental plates. Second by second the years will go by, by the thousands, by the millions. And the sun will continue to mark the time, rising and setting every day, a constant in a changing world. Or so it will appear. In effect, the sun, too, is constantly changing, not only in its superficial manifestations, more or less periodic, but above all in its interior.

THE GREAT HEAT WAVE

Every day, part of the hydrogen deep in the sun's interior turns into helium and a part of the sun's mass is irretrievably lost, transformed into energy that is dissipated in space. This is the price the sun has to pay to keep itself in equilibrium, that is, to remain the same, at least outwardly, to keep the same diameter, and to radiate the same amount of energy. However, this is not entirely true.

Hydrogen began to burn in the central region of the sun about 4.7 billion years ago. Since then, as a helium core formed, hydrogen has burned in a fairly thick shell farther and farther away from the center. The rate of hydrogen burning has accelerated but the surface temperature of the sun has remained the same; its diameter, and hence its luminosity, have increased. Altogether, the sun will spend over 10 billion years on the main sequence; 4.7 of them have gone by, and in another 4.5 the sun's luminosity will have increased by 50% and its diameter by 25%.

As long as the sun's variations are small and slow they do not have serious consequences for the earth, but 10 billion years after the sun has settled on the main sequence radical changes will begin to occur in its interior. By this time all the hydrogen in the central region will have turned into helium, and hydrogen burning will come to a stop. The helium core will contract, generating energy that will partly raise the core's temperature and will partly heat up the overlying layer of unburnt hydrogen. Nuclear reactions will start there, but only within a thin shell. In the process, so much energy will be generated that the upper layers will expand. The sun will rapidly turn into a red giant and continue to grow larger and brighter for the next billion years. Eventually it will become 100 times larger and 1,000 times more luminous than it is today.

Although its temperature will be somewhat less than today's, the sun will be so large that it will burn the earth with a heat 1,000 times greater than the current amount. Seen from the distance of the earth, the sun will have an apparent diameter of 60°. At noon its huge disk, dull red and terribly hot, will cover a third of the entire sky. The spectacle should be awesome, especially if seen from the regions between the equator and the tropics, where the sun passes through the zenith: a huge red disk covering a large part of the sky. But no one will be able to see this view from the earth's surface because the oceans will have evaporated, blanketing the earth with a dense atmosphere, unless the heat has dissipated it into space. In the feeble reddish light that will filter down, continental plains, mountain chains, and ocean basins will lie equally scorched and lifeless. No one will be there to behold this hellish vision because life will have long since gone, obliterated by the heat wave that engulfed the earth when the sun turned into a red giant.

What will happen next? We already know from stellar evolution. The sun will go through the helium flash, a brief return to a size and luminosity closer to the current ones, another expansion into a red giant, and then the final decline to a white dwarf. We have learned what the ultimate fate of the sun will be and how the history of life on earth will come to an end with the first great heat waves from a body that will no longer be "our" sun.

IN THE GRIP OF ICE

The earth is now a frozen planet. Whatever is left of the atmosphere has solidified, covering the soil with an icy mantle that levels every feature. Buried beneath the ice is every vestige that life has ever left, from the imprints that the primitive fauna of Ediacara left on the ocean floor to the works of *Homo sapiens* and his successors, those species that at present are hidden in the mists of the future and are to be things of the past. The sun still rises on this frozen world: a bright dot in a black starry sky (because devoid of atmosphere) that still sends much light, a ghostly white light that although 400 times fainter than the sunlight of today is 600 times more intense than the silvery light of the full moon. The planets also rise and set (except for Mercury, which almost certainly has been destroyed by the sun swelling into a red giant). But they are all invisible because the sunlight they receive and reflect is very faint. Even the planets that were too far to be damaged by the sun's expansion are now inert blocks in the grip of everlasting cold, poor relics where even the hope of life has failed. This is what the solar system will be like for billions and billions of years after the history of the earth and the other planets has come to an end.

Does this mean that the evolution of the only planet known to have a living cargo will be nothing but a long journey from birth to death, like the fate of all its inhabitants? It would appear so. But before we start crying over our present and future tombs, we should consider one last fact.

We have witnessed a marvelous evolutionary process that began on the earth nearly 4 billion years ago and thus far has culminated in mankind. If the earth is the only place in the universe where such a process has occurred and evolved, then we know that one day there will be an end to it and the universe once more will become an aggregate of lifeless matter, a stupendous entity with no one there to be aware of its existence. It might happen soon if the human race is wiped out by a nuclear war or an ecological disaster. Or it may happen in a few billion years when the expanding sun burns the surface of the earth to a cinder. If that will be the end, there remain 5 billion years during which ever more intelligent species can develop that may be able to save themselves by escaping to other planetary systems. Considering the large distances involved and the need to adapt rapidly to habitats quite different from the environments to which man and his intelligent successors had become accustomed over millions of years, it seems more likely that biological evolution will be restricted to our immediate vicinity and perhaps to the earth itself. If so, when terrestrial life forms die out, this complex matter capable of producing intelligence, feelings, intuition, and religious ardor could disappear forever from the universe.

But there is one possibility that this will not happen—if the process of life that developed on earth is not unique; if matter evolves elsewhere as it did on our planet: in brief, if there is life also in other parts of the cosmos.

We have seen how life began and evolved here on earth starting from inorganic matter. Now we ask, Is this process unique or could it occur elsewhere? Is it a chance occurrence and hence very rare, or is it part of the natural evolution of matter throughout the cosmos?

Twenty years of advances in our knowledge of the planets, stars, and interstellar medium allow us to address these questions on sounder scientific bases than were previously available. The results of our research into extraterrestrial life may be summarized as follows: On the one hand, space probes have found no sign of life either where we were most hoping to find it (Mars) or on any of the other planets and satellites explored to date; on the other hand, radioastronomy and the study of meteorites have revealed the presence of prebiotic molecules in space. Let us proceed to examine the facts and their main interpretations.

PRELUDE IN SPACE

In many regions of space, particularly in the band of the Milky Way, the telescope shows large, bright nebulae that to the eye of the observer generally look evanescent, as if they were mists about to dissolve. But after long exposures the photographic plate shows them denser and more extended, with a richer palette of colors and shades. Interspersed among these bright nebulae, or isolated in space, there are dark nebulae that can be seen because they stand out against the luminous background of bright nebulae or star fields. Bright and dark nebulae alike consist of gas and dust; the only difference between them is that the former are illuminated or excited by nearby stars, while the latter are not. It is in these regions and from this material that new stars and planets are formed.[40]

It was known for a long time that the gas in these nebulae consisted of various elements, including oxygen, nitrogen, sulfur, and above all hydrogen. But no compounds were known until two diatomic molecules, CH and CN (cyanogen), were discovered from spectroscopic observations in 1937 and 1939, respectively. In 1963 radiotelescopes detected another molecule, the hydroxyl radical OH, important chemically because it forms water when an atom of hydrogen is added to it. The detection of the hydroxyl radical in diffuse nebulae was followed a few years later by the discovery that it also could be concentrated in extremely compact regions. In January 1968 it was found that the OH emission from a radiosource in Cassiopeia actually came from seven discrete sources with angular diameters not exceeding .02 second of arc. Shortly thereafter, two radiotelescopes, one in the United States and one in Sweden, were linked so as to constitute one huge instrument with a diameter equivalent (with regard to resolving power) to 7,720 km, which is the distance between the two telescopes. Measurements with this instrument showed that the apparent diameter of the smallest source was only .005 second of arc. In linear dimensions, taking into account the source's distance from us, this corresponds to the diameter of the orbit of Mars. Hence the emission must originate from a very compact region, perhaps a star.

Subsequently more compact OH emitters were detected. One coincided with the infrared B–N source discovered in Orion in 1967, and others, with infrared stars or regions of ionized hydrogen. The unexpected feature of all these radio sources is that a considerable amount of energy is emitted within a relatively small volume by a process of amplification suggestive of maser action. (Maser stands for *m*icrowave *a*mplification by *s*timulated *e*mission of *r*adiation.)

The year 1968 also saw a new surge in the discovery of interstellar molecules, beginning with the detection of ammonia (NH_3). Early in 1969 astronomers discovered the water molecule (H_2O), whose presence, given the abundance of hydrogen in the cosmos, had been suspected since the detection of hydroxyl. In March 1969 they found formaldehyde (H_2CO), the first organic polyatomic molecule to be observed.

In 1970 B. F. Burke and his collaborators found that some H_2O radio sources had very small angular diameters. Subsequent research showed that H_2O also could emit radiation by a maser-type mechanism, as had been found previously for the radical OH. Furthermore, the two emitters often were associated; that is, they were present in the same object. Sometimes they are observed in red giants of late spectral type, which, if you recall, are stars approaching the final stage in their life cycle. More often they are found in regions where stars are forming. In this latter case the OH and H_2O masers are close to strong infrared sources, presumably protostars.

There are reasons to believe that the phenomenon of maser emission occurs at the beginning of a star's life and does not last very long. In any case, the H_2O masers appear to vary in intensity with the passage of time. In May 1977 A. D. Haschick and B. F. Burke were the first to witness the onset of this phenomenon in the region known as W 3. On 8 May they observed an increase in the intensity of the H_2O line at the wavelength of 1.35 cm, which lasted until 17 May and was followed by a slow decline over the next 4 weeks. Almost certainly it was an H_2O maser located at the periphery of the dust cocoon surrounding a very young O5 star, and associated with a temporary escape of energy lasting a couple of days.

As to the interstellar molecules, many more of them have been discovered since 1970, all but one, the hydrogen molecule (H_2), from their radio-frequency emission. They include hydrocyanic acid, methyl alcohol, ethyl alcohol, and hydrosulfuric acid. Particularly important was the discovery of methylamine in 1974, because it can react with formic acid to form glycine, the simplest of the amino acids that combine to form proteins. As you know, proteins are essential constituents of all living cells. Announced in February 1975 was the discovery of acrylonitrile, a molecule composed of 7 atoms and belonging to the group of the olefins, which are chemicals we use to manufacture synthetic fibers and rubber and some plastic materials. At present over 50 types of molecules are known, some of which are more complex than acrylonitrile (see table 5). The largest and heaviest found so far is HC_9N, with 11 atoms.

Two very important facts emerge from these findings. In the first place, in interstellar space there are molecules closely connected with the organic world. Until a few years ago even the strongest advocates of extraterrestrial life stopped short of claiming that it could be similar to our own. They assumed that there would be quite different forms of life—for example, organisms based on the chemistry of silicon rather than carbon. Now we know that in interstellar space, wherever the temperature is low enough to permit the formation of molecules, there form precisely those compounds like water and ammonia that are closely related to the type of life found on our planet, and even some compounds such as formaldehyde, formic acid, and methyl alcohol, that are part of our organic chemistry.

In the second place, most of these molecules are concentrated in relatively small spaces, often in regions like the Orion nebula where stars are born. Thus the basic molecules of our organic world are scattered

Table 5 Molecules discovered in interstellar space up to 1980[a]

Name	Chemical formula	Year of discovery	Name	Chemical formula	Year of discovery
Methylidine*	CH	1937	Methylamine	CH_3NH_2	1974
Cyanogen*	CN	1940	Dimethylether	CH_3CH_3O	1974
Hydroxyl*	OH	1963	Ethyl alcohol (ethanol)	CH_3CH_2O	1974
Ammonia	NH_3	1968	Sulfur dioxide	SO_2	1975
Water*	H_2O	1968	Silicon sulfide	SiS	1975
Formaldehyde	H_2CO	1969	Vinyl cyanide (acrylonitrile)	H_2CCHCN	1975
Carbon monoxide*	CO	1970	Methyl formate	$HCOOCH_3$	1975
Hydrocyanic acid*	HCN	1970	Nitrogen sulfide	NS	1975
Cyanoacetylene	HC_3N	1970	Cyanamide	NH_2CN	1975
Hydrogen	H_2	1970	Cyanodiacetylene	HC_5N	1976
Methyl alcohol	CH_3OH	1970	Formyl	HCO	1976
Formic acid	HCOOH	1970	Acetylene	C_2H_2	1976
Formamide	$HCONH_2$	1971	Nitroxyl	HNO	1977
Carbon sulfide*	CS	1971	Ketene	C_2H_2O	1977
Silicon oxide	SiO	1971	Propionitrile (ethyl cyanide)	CH_3CH_2CN	1977
Carbonyl sulfide	OCS	1971	Carbon*	C_2	1977
Methyl cyanide (acetonitrile)*	CH_3CN	1971	Cyanooctatetryne	HC_9N	1978
Isocyanic acid	HNCO	1971	Methane	CH_4	1978
Methylacetylene	CH_3CCH	1971	Nitrogen oxide	NO	1978
Acetaldehyde	CH_3CHO	1971	Butadienyl	C_4H	1978
Thioformaldehyde	H_2CS	1971	Hexatricyanogen	HC_7N	1978
Isocyanidric acid	HNC	1971	Methanethiol	CH_3SH	1979
Hydrosulfuric acid	H_2S	1972	Isothiocyanic acid	HNCS	1979
Methanamine	H_2CNH	1972	Cyanoethynyl	C_3N	1980
Sulfur monoxide	SO	1973			
Ethynyl	C_2H	1974			

a. Many of the molecules discovered in space are radicals. Some of those listed in the table have also been found ionized. Others have been found with deuterium atoms in place of one or more hydrogen atoms; these molecules have not been included in the table. Molecules marked with an asterisk have also been found in comets.

throughout the universe, and they form where the stars are formed. This fact is particularly important for their survival because if they floated free, they would soon be destroyed by the x-rays, cosmic rays, and ultraviolet radiation that pervade space. Inside the nebulae, instead, they are mixed with dust grains that can incorporate and protect them. When the dust condenses to form a star and its vast protoplanetary cloud, the molecules become part of the new planetary system, where they will be able to develop further. This hypothesis is supported by another great discovery of the 1960s and early 1970s.

In this period a number of special meteorites known as carbonaceous chondrites were found to contain prebiotic molecules such as fatty acids, purines, and pyrimidines. Unfortunately, since the specimens had fallen between 1806 and 1965, terrestrial contamination could not be ruled out. But as luck would have it, on 28 September 1969 a large meteorite exploded over the town of Murchison in Australia, strewing rocks over an area 8 km long and 1.5 km wide. The meteorite was a carbonaceous chrondrite of class II, an age group between 4.5 and 4.7 billion years. Hence the body had formed earlier than the earth's crust but not earlier than the solar system and could give us information about that remote and crucial epoch. Examining the remnants of this meteorite and a similar one that had fallen at Murray, Australia, on 20 September 1950, C. Ponnamperuma and his coworkers found that they contained 18 amino acids and 2 pyrimidines, the very same ones in both meteorites. This means that if the two meteorites did not happen to be fragments of the same body but came from different regions of space, perhaps distant, those amino acids were widely produced in the early days of the solar system. Of the 18 amino acids, 6 are of the type normally found in proteins, namely, aspartic acid, glutamic acid, glycine, alanine, α-aminoisobutyric acid, and β-alanine. The other 12 are similar to them but play no role in terrestrial living matter. Nor are the two pyrimidines of the type found in terrestrial organisms. This indicates that the molecules found in the meteorites were not due to terrestrial contamination. Thus we can confidently state that in space, besides a large amount of organic molecules, can be found the very prebiotic molecules that are built from them, such as amino acids and pyrimidines, which, in turn, form the macromolecules of life: proteins, DNA, RNA, polysaccharides, and lipids.

Although we have not found life outside the earth, we have found what we know to be the link between inorganic and organic matter here on earth. Considering the great abundance and variety of organic molecules discovered in only ten years and the fact that when interstellar matter condenses into a planetary system some of the organic molecules contribute to the formation of prebiotic molecules, one is tempted to conclude that such molecules may well be the link between inanimate and living matter everywhere in the cosmos. In other words, life may always be based on the chemistry of carbon as it is here on earth.

When you think about it, this conclusion is fairly reasonable. If we exclude helium and neon, which do not combine with other elements, the most abundant elements in the universe are hydrogen, carbon, oxygen, and nitrogen, that is, those that constitute the basis of organic chemistry and 96% of our bodies.

Aside from the hydrogen and oxygen in water and in the atmosphere, the most abundant elements on the earth are iron, silicon, nickel, and magnesium. Clearly, when we ignore the sky the composition of the human body seems very peculiar. In fact, it is peculiar only with respect to the planet earth. When we analyze the rest of the universe we find that the earth, not our body, is the exception to the norm. The stark evidence is in table 6. It shows that in a sense we are children of the cosmos more than of the earth. Nonetheless, as the product of billions of years of evolution, we have taken from our planet some twenty additional elements. Although they constitute a very small percentage of our bodily makeup (some are only in trace amounts), we could not live without them. The most important of these elements are fluorine, sodium, magnesium, phosphorus, sulfur, chlorine, potassium, calcium, manganese, iron, cobalt, copper, zinc, molybdenum, and iodine.

In conclusion, it appears that life may have its prelude in space, starting from the most abundant elements. Subsequently life would develop in the many varieties we have seen on the earth and perhaps in a myriad of alternative forms on the countless worlds scattered throughout the cosmos. Each different form of life may be determined by the type of material that predominates on the celestial body where it develops; the trace elements essential to our bodies may differ from planet to planet. Moreover, chance may play a role by producing different organic molecules, like the 12 amino acids and the 2 pyrimidines found

Table 6 Distribution of the most abundant elements in the sun, in the earth, and in living organisms (%)

Sun		Earth		Earth's crust		Earth's atmosphere		Bacteria		Human beings	
Hydrogen	93.4	Oxygen	50	Oxygen	47	Nitrogen	78	Hydrogen	63	Hydrogen	61
Helium	6.5	Iron	17	Silicon	28	Oxygen	21	Oxygen	29	Oxygen	26
Oxygen	0.06	Silicon	14	Aluminum	8.1	Argon	0.93	Carbon	6.4	Carbon	10.5
Carbon	0.03	Magnesium	14	Iron	5.0	Carbon	0.011	Nitrogen	1.4	Nitrogen	2.4
Nitrogen	0.011	Sulfur	1.6	Calcium	3.6	Neon	0.0018	Phosphorus	0.12	Calcium	0.23
Neon	0.010	Nickel	1.1	Sodium	2.8	Helium	0.00052	Sulfur	0.06	Phosphorus	0.13
Magnesium	0.003	Aluminum	1.1	Potassium	2.6					Sulfur	0.13
Silicon	0.003	Calcium	0.74	Magnesium	2.1						
Iron	0.002	Sodium	0.66	Titanium	0.44						
Sulfur	0.001	Chromium	0.13	Hydrogen	0.14						
Argon	0.0003	Phosphorus	0.08	Phosphorus	0.10						
Aluminum	0.0002			Manganese	0.10						
Calcium	0.0002			Fluorine	0.063						
Sodium	0.0002			Strontium	0.038						
Nickel	0.0001			Sulfur	0.026						
Chromium	0.00004										
Phosphorus	0.00003										

in the Murchison and Murray meteorites. These molecules, and many others unknown on earth, might very well manufacture macromolecules for other forms of life, different from ours, but like ours in their being based on carbon chemistry.

WHERE LIFE IS BORN

The great abundance of organic molecules in space, particularly in the regions where the stars and the planets are formed, and the fact that prebiotic molecules such as amino acids and pyrimidines can be incorporated into the material that condenses into a planetary system suggest that the origin of life should be considered in the wider context of the birth of a star and its planetary system. Maybe the organic molecules in the protoplanetary cloud were partially or entirely destroyed during the formation of the planets because of, among other things, the intense heat generated in the process. But maybe not. After all, the amino acids found in the meteorites have withstood equally severe tests. Furthermore, conditions even more favorable to their formation could have recurred on each planet after the formation of a first solid crust and an atmosphere.

What is still mysterious, however, is the transition from prebiotic molecules to the self-replicating biological macromolecules that formed the first cells. As we have noted, this transition may be the result of a still unknown process. The extraordinary thing about it is that this process must be fairly simple. On the earth the entire transition from inorganic matter to the first cells would have taken 300–350 million years, while the evolution of the first procaryotes into eucaryotes would have required another 2,300 million years, about 6 times longer. It may be that the disproportion between the two transition times appears inexplicable only to our eyes. Perhaps we think that the first transition should have taken longer because, not knowing how it did occur, we attribute it to chance, which requires long times. The second transition, which we would have expected to be more rapid, perhaps happened more slowly because for millions of years there was no necessity for procaryotes to evolve into eucaryotes.

But there may be another explanation, namely, that life did not originate on the earth but came from somewhere else. In other words, not only organic and prebiotic molecules but life itself, in its most primitive forms, might originate in space. This hypothesis, first advanced by the Swedish chemist S. Arrhenius in 1908, has been proposed anew by F. Hoyle and C. Wickramasinghe in a recently formulated theory. They hold that in the same nebulae where so many organic molecules have just been discovered, there also form various biochemical units such as porphyrins, some bases of nucleic acids, some amino acids, and particularly polysaccharides, whose presence they believe to be evidenced by an absorption band observed in the infrared. All this material would be scattered in the nebulae, mixed with the other matter from which stars are born. According to these authors, the kindling of a star destroys the prebiotic material in its vicinity, but the rest of it is blown away by radiation pressure toward regions where other stars may be forming. Thus an aggregate of molecules may be pushed back and forth many times among the denser clouds in which stars are born. As the molecules come and go across regions of very different physical conditions, a sort of Darwinian evolution would occur, resulting in the emergence of molecular structures more resistant to the hostile environment of the interstellar medium.

According to this theory, when the solar system was born the prebiotic molecules contained in the material that formed the inner planets (Mercury, Venus, the earth, and Mars) were destroyed, and no primitive atmosphere formed around the earth from the gases released by the cooling rocks. However, prebiotic molecules survived at the outer edge of the solar system and were incorporated into blocks of frozen material, where they eventually produced primitive forms of life. These blocks of ice are none other than the comets whose orbits, at first very large and nearly circular, in time become elongated ellipses passing by the sun and the inner planets.

The theory also maintains that when the comets began to pass near the sun, some fell to the earth, still airless and lifeless, and their material produced an atmosphere that softened the landing of the next comets and made it possible for the first procaryotic organisms to reach the earth's soil without being destroyed by the heat of a violent impact.

This theory is thus in agreement with Oort's concept of a comet reservoir located about halfway between the sun and the nearest star.[41] One should also keep in mind that the entire solar system travels in space, including the swarm of comets surrounding it. According to Hoyle and Wickramasinghe, as the solar system passes through the

spiral arms or other regions rich in gas and dust, new prebiotic molecules are swept up and stored in the comet reservoir. Every time one of
these comets leaves the distant swarm for the sun's neighborhood it
seeds interplanetary space and planets with fragments of material containing living matter. By this process, still going on today, space-born
viruses also would be brought to earth—in particular, the viruses that
cause the common and widespread flu epidemics. This aspect of the
theory is somewhat farfetched and has raised a veritable outcry from
biologists and epidemiologists. Furthermore, the presence of polysaccharides in space is very much in question. It all revolves on the interpretation of an absorption band observed in the infrared. Hoyle and
Wickramasinghe attribute it to polysaccharides, but their assumption
has been criticized by W. G. Egan and T. Hilgeman, who support the
accepted view, namely, that the band is produced by two distinct components, ice and silicates. Of course, this does not preclude that macromolecules may exist in space, but at present there does not seem to
be any observational evidence for them. In any event, even if things do
not go exactly as Hoyle and Wickramasinghe have theorized, their
main idea might still be valid: Life might begin in space and then float
down to existing celestial bodies. With regard to the planet earth, this
would explain why a relatively short time elapsed between the formation of the crust and oceans and the appearance of the first life forms.

DEVELOPMENT

If life originated in space from the four most abundant elements in the
universe, it could be a normal process, and hence quite common. Its
further development, or lack of it, would then depend solely on the
conditions of the celestial bodies upon which the "life principles" happened to fall. Let us see what sort of place would be suited to the development of life forms with the characteristics and structures I have
described.

Pervaded as it is by lethal radiation, interstellar space can be immediately ruled out. The organic molecules born there can only give rise
to macromolecules and perhaps to primitive organisms that at most
survive by being encapsulated in dust grains.

Suitable places for life to flourish would certainly be planets (or satellites) with a solid crust, oceans, and an atmosphere. As we have seen

here on earth, oceans and atmosphere are protective elements that favor the survival of primitive life forms. Besides protection, evolution needs an active planet, that is, a planet subject to changes, preferably slow ones; otherwise the life developing there might not have time to adapt and could undergo a catastrophic arrest.

But a planet must be more than just hospitable and active. For life to take hold, the planet must receive the right amount of light and heat from its sun. In other words, it must be neither too hot nor too cold nor subject to large changes in temperature. Hence its orbit must be nearly circular, and its average distance from the source of heat and light must be the greater the hotter and brighter the source is. Active life can exist between certain extremes of temperature, which here on earth are roughly −60 to +70°C. Thus there are not one but many possible orbits in the space delimited by two concentric spheres whose common center is the sun, around which the hypothetical planet suitable for life circulates, and whose radii correspond to the distances that provide the maximum and minimum tolerable temperatures. This region, known as an ecosphere, can comprise the orbits of various habitable planets. On the other hand, given the same amount of heat from its sun, the surface temperature of a planet may vary depending on the type of atmosphere surrounding it. For instance, an atmosphere that because of its physical and chemical properties keeps out or retains the right amount of heat can maintain a temperature suited to life even on a planet outside the bounds of the ecosphere.

Another important condition is that the star of our hypothetical planet should be stable for a sufficiently long time; otherwise life would not have time to develop or even take hold. This rules out type-O stars because they evolve in a few million years, changing rapidly in intensity and ending in a supernova explosion that burns their planets to cinders. We can also exclude B stars and many A stars since they remain on the main sequence for a time ranging from 8 million years for B0 stars to 2 billion years for A5 stars, which is still too short. From the point of view of having habitable planets, the most suitable stars are the smaller ones, like the sun or even smaller, which live billions of years. Along with K and M dwarfs, these stars are very numerous, and they can have life-sustaining planets even if their surface temperatures are much lower than the sun's. They emit less energy and heat, but some planets can be well heated simply by being closer to them. When the fire is low, one can still feel warm by sitting near the fireplace. On the other hand, these faint stars live much longer than the sun, and their planets could harbor strange beings almost blind to our light but well able to see in the red and infrared light with which their sun "illuminates" their worlds.

Another condition essential for life is that the radiation received by a planet should be constant or vary slowly over long periods of time. Hence the sun of our planet cannot be a variable star. But this condition is not too restrictive because there are no stars that vary throughout their entire lives. Variability is a phenomenon that occurs in particular periods of a star's life—for example, in the initial stage—but at that time the planets are just forming. Another critical phase comes late in a star's life, when it leaves the main sequence and swells into a red giant. At that point it will destroy the inner planets, with or without inhabitants, as well as the life, if any, on the outer planets.

A planet can also be subject to large changes in temperature when it travels an unstable orbit in a system of two (or more) stars. In this case, the planet may receive too little heat to sustain life when its orbit takes it very far from both stars; just the right amount of heat, for a short time, as it approaches the stars; and much too much heat when it passes close to them. At that point, if it is a close binary system, the planet might even be engulfed by the fiery material that surrounds the stars or flows from one onto the other. For this reason, astronomers in the past had excluded all double or multiple systems—about half of all existing stars—as the type that could have habitable planets. In 1977, however, R. S. Harrington reconsidered the problem with the help of modern computers and found that in fact there can be stable planetary orbits in a binary system. More precisely, he defined two types of orbits, one circling one of the stars (figure 69a) and the other circling both stars (figure 69b). Naturally, even in such systems not all orbits are stable. In the first case, to have a stable orbit the distance between the stars must be 3.5 times greater than the distance between the planet and the inner star. In the second case, the distance of the planet from the center of mass must be more than 3.5 times the distance between the stars.

We can use the solar system as an example. If Jupiter were a star like the sun, mass included, life on the earth would be more or less the same. Our orbit around the sun would be stable. There would be

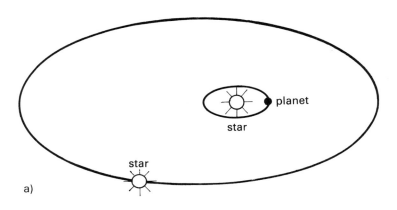

a)

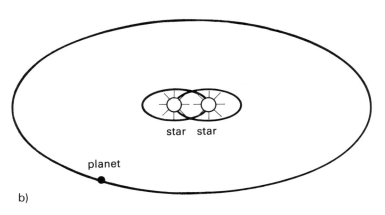

b)

Figure 69   Diagrams illustrating stable planetary orbits in binary systems: (a) planet orbiting one of the stars; (b) planet orbiting both stars. [From *Mercury*, © 1978 Astronomical Society of the Pacific]

two suns, but our temperature would not be much higher because the second sun (Jupiter), due to its distance, would appear smaller and much weaker. At times we would see our two suns close to each other, one large and bright as usual, its twin looking one-fourth smaller and much fainter; after the suns set, night would fall just as it does today. At other times of the year the second sun would be somewhat larger and brighter; it would rise at sunset and set at sunrise, and during that time night would not come to the earth. There would be 24 hours of daylight, 12 brighter and 12 less bright.

The orbit of the earth could remain stable also in the second case (figure 69b)—for example, if we replaced the sun with a close binary system whose components were at an average distance from each other of about 30 million km. We would simply orbit both suns, observing their relative motions and their different eclipses as various portions of one or the other star are occulted and the stars keep on changing roles, now eclipsing and now being eclipsed. If the orbital plane of the earth happened to be somewhat inclined to that of the two stars, as we revolved about them, sometimes we would see them aligned, sometimes, so to speak, a little from above, and sometimes a little from below. And the continuous combining of these circumstances in different ways would cause these phenomena to occur with a periodicity so diluted in time as to show a different configuration every day. We are not in these conditions because we have only one sun, but according to Harrington's calculations in our Galaxy there can be earthlike planets in stable orbits around a double star. And these planets might have inhabitants who normally witness what we can barely envision in our minds.

TENTATIVE ESTIMATES

At present the only life we know is here on the earth. But despite the fact that the search for extraterrestrial life on our closest neighbors, Mars and the moon, has failed so far, what precedes life seems to be a widespread characteristic of highly evolved cosmic matter, provided its temperature is not too high. Taking into account all the conditions mentioned above, it should be possible to estimate the number of life-bearing planets that exist in the Galaxy and in the whole universe.

This is easy in principle but not in practice because much of the data is a matter of opinion, and therefore different authors arrive at different results. The number we are seeking (N) is given by a very simple expression,

$$N = \frac{n}{A} \times f_p \times P \times p_l \times D_l,$$

where

n is the number of stars in the Galaxy capable of furnishing their planets with enough energy to sustain life,
l stands for life,
A is the age of the Galaxy expressed in years,
n/A is the number of suitable stars that form every year in the Galaxy,
f_p is the fraction of stars that have planetary systems,
P is the number of habitable planets that on the average exist in every planetary system,
p_l is the probability that life actually will develop on a habitable planet, and
D_l is the duration of life on a planet (expressed in years).

To find N we only have to replace each symbol with the appropriate number. If we do not multiply by D_l, we obtain the number of habitable planets that form every year in the Galaxy. If we multiply the result by D_1 we find the number of planets in the Galaxy that could be currently inhabited. The problem in using this formula is that none of the numbers to be inserted is accurately known. In some cases—for example, the probability that life actually will develop on a suitable planet—we have only opinions, more or less well-founded. Nevertheless, we can attempt to make a cautious and not overly optimistic estimate.

For the purpose of this calculation, the number of stars in the Galaxy, n, is generally assumed to be 10^{11} (100 billion). The actual number of stars certainly is higher, at least 2×10^{11}, and according to the latest (1981) results obtained by B. J. Bok there could be as many as 10^{12}, or 1,000 billion stars. However, since many stars do not live long enough to have life-sustaining planets, it seems reasonable, in-fact somewhat conservative, to take 100 billion as the number of stars suited to supporting life in their neighborhoods. In the past f_p was set at 0.5 on the assumption that all except double or multiple stars (estimated at about half the total number) have planetary systems. From Harrington's recent calculations, however, we can assume that a number of double stars also have planets in stable orbits. Thus it may be reasonable[42] to set $f_p = 0.7$.

To estimate the number of habitable planets in a planetary system we have only the solar system to go by. Since at present only the earth appears to be habitable, many authors assume $P = 1$. In effect, except for soil conditions, at least one other planet, Mars, has characteristics fairly similar to the earth's. Along with Venus and some satellites of Jupiter and Saturn, Mars might have been habitable in the past, or might be so in the future, possibly for short periods of time and for primitive forms of life. At this stage of our knowledge, however, it is safer to exclude any case that may represent a borderline situation, and assume $P = 1$.

As we have seen, today we are inclined to believe that whenever possible, life did develop, whether it originated from the prebiotic material incorporated in the original matter of a planetary system or it arrived on suitable planets from interstellar clouds. Thus we shall assume $p_l = 1$. Let us remember, however, that there is no evidence for this assumption, only clues. Finally, we can set the age of the Galaxy as $A = 10^{10}$ (10 billion) years, which is certainly less than its actual age.

Once we have made our calculations, we find that for every year in the life of the Galaxy 10 stars are born that could support life on their planets and that there are 7 cases per year in which life should actually develop.

As we have seen, a planet can harbor life only as long as its sun is on the main sequence. This period lasts at least 10 billion years for sun-like stars and much longer for less massive stars; hence we shall set $D_l = 10^{10}$. Thus the duration of life on a planet coincides, obviously by chance, with the age of the Galaxy. To obtain the number of life-bearing planets currently in the Galaxy, we should now multiply 7 (the number of planets where life begins every year) by 10 billion years (the assumed duration of life on a planet). Of course, life would be nearing extinction on planets as old as the Galaxy, while it would be just beginning on planets of recent formation and in the first stages of development—those that on earth took place 2–3 billion years ago.

However, there is a difficulty that forces us to revise our numbers. We do not know whether the processes that pave the way for the development of life have been in effect since the birth of the Galaxy. Hence it would be imprudent to obtain the number of planets that currently harbor life by multiplying the number of planets where life begins every year by 10 billion years. But there is no doubt that such processes were in effect when the earth was formed. Thus, to be safe, we shall multiply 7 by 5 billion years, which is a little more than the age of the solar system and half of the 10 billion years assumed to be the duration of life on a planet. In this manner we find that at present there should be 35 billion life-bearing planets in our Galaxy. Based on this number and estimating that there are at least 10 billion galaxies, it turns out that there would be 350 billion billion life-bearing planets in the universe. It is quite an impressive figure, even granting that it could be any kind of life, even primitive, and not necessarily intelligent.

Furthermore, there are reasons to believe that this number should increase considerably with time because of the life-bearing planets belonging to stars of late spectral types. As you recall, these stars remain on the main sequence far longer than the sun; they can illuminate and heat their planets, weakly but steadily, for tens of billions of years.

Continuing in this vein, the next obvious question is this: Among the forms of life that are born every year in our Galaxy, how many would evolve into intelligent species, and how many of the latter would develop a technological civilization like ours? To answer this question we would have to introduce into our formula two more probabilities, p_1 and p_2, corresponding to the first condition and the second condition, respectively. In addition, in order to estimate the number of technological civilizations that currently might exist in our Galaxy we would have to multiply the new result by the duration of a technological civilization rather than by D_1, the duration of any form of life.

Here we truly begin to wander in the dark. Our only reference is the history of the earth. That is not enough, and in any case we do not know when or how our civilization will come to an end. All we know from experience is that once an intelligent species emerges, technological development is very rapid and its very achievement provides the civilization that has produced it with powerful means of destruction. It is fair to say that a civilization unable to control its own technology can develop and self-destruct in times on the order of 100 years. If it can overcome the crisis by achieving a balance between production and consumption, it may last for an indefinite time, which could be interrupted only by other causes, such as the emergence of stronger species, planetary catastrophes, or cosmic events. The duration of a technological society that has overcome the 100-year crisis may be considerable—

for example, on the order of 10 million years. This may be an optimistic estimate in that we are familiar only with the 100-year crisis and there may be other threats to a technological civilization even after it has stabilized. In other words, during those 10 million years the 100-year crisis may be merely the first, not the only one.

Of course, it would be the only crisis for any civilization that cannot overcome it. But there is one more thing to take into account. If a society of intelligent beings creates a technological civilization that self-destructs, it is possible that after a time a new intelligent species will evolve on that planet. If the suicidal society has not used up all the natural resources of the planet, the new beings might develop another technological civilization. In the worst case, to the extent that life would start not from zero but from surviving species capable of becoming intelligent, there might arise an intelligent but nontechnological society that potentially could last much longer. In other words, it would be a civilization more like that of classical Greece than our own. Though such a civilization might be capable of great achievements, it could never build the devices necessary for traveling, sending, or receiving messages across space, and therefore it would become isolated. Perhaps there are many such civilizations in the universe, either by necessity or by choice.

LIFE IN THE UNIVERSE

We have seen that in our universe inanimate matter evolves in such a way as to produce life; we have tried to estimate the number of planets that harbor life; we have reasoned, computed, and fantasized, but we have not found life outside the earth.

This is not really surprising. We had already assumed, based on our recent explorations of the solar system, that similar planetary systems would contain only one life-bearing planet. Also, we have estimated that at present there should be 35 billion life-bearing planets in our Galaxy. This means that for every three stars whose surroundings we explore we should find only one with a life-bearing planet. Thus far we have explored, and poorly at that, the neighborhood of only one star, the sun.

It is hoped that we shall continue to search for extraterrestrial life in the near future, either directly, with manned spacecrafts or automated probes, or indirectly, by sending radio messages and monitoring the sky for signals sent by other technological civilizations like our own, though not necessarily sent to us.

But this is not our main concern at the moment. As we attempt to reconstruct the history of the universe in all its aspects, including living matter, we are not so much interested in where extraterrestrial life could be as in whether it exists and what it would be like. As for the solution of the first problem we have already found much evidence in favor of an affirmative answer. Thus let us move on to the second question, which can be explicitly expressed in these terms: What could life be like and look like on other worlds?

The physical appearance of extraterrestrials is something that has teased our imagination far more than anything else about them, including their very existence. Indeed, we have always been more interested in portraying than in finding them, and naturally we have imagined them to be similar to the life forms we know, particularly ourselves. Of course, it would be much more important to establish communications with beings from another world, which could reveal to us mysteries and problems they have already solved. This could happen if we ever found creatures more advanced than we are—not too advanced, however, or we would find ourselves in the situation of a man who tries to explain differential calculus to a hen, we being the hen in this case. But the curiosity, the wish to see them would always remain, even at the cost of being repelled by their monstrous appearance. On the other hand, we might not be repelled at all. They might look beautiful, though different from us. They might even be similar to us or to some terrestrial animals.

This last case is by no means improbable. On the contrary, it seems quite likely. Judging from the type of molecules found in interstellar space, extraterrestrials should be made of the same material as ourselves. We also have assumed that they have developed in environments physically and chemically similar to our own. These two factors could put in motion the process known as evolutionary convergence, whereby starting from entirely different origins there develop similar structures and shapes, required by similar environments and modes of life. A classic example is a sensory organ like the eye. Since there is light and every creature needs to see, the eye has developed in a similar way in such diverse animals as fish and snakes, humans and insects.

dolphin

ichthyosaur

shark

Figure 70   An example of evolutionary convergence: the dolphin, the ichthyosaur, and the shark. Although they do not descend from common ancestors, these animals have similar shapes and structures because of their adaptation to the same environment—in this case, the ocean. [From *Il pianeta dell'uomo*, vol. VI, Mondadori; illustration by F. Ghiringhelli, Milan]

Another and even better example is given by the comparison of three aquatic animals—the extinct ichthyosaur, the porpoise, and the shark—which are remarkably similar in appearance though being, respectively, a sauria, a mammal, and a fish (figure 70). A third example is the parallel development of the North American and South American fauna of the Miocene, before the two continents were linked by the Isthmus of Panama. While a fauna of placental mammals was evolving on the northern continent, the marsupials were independently evolving on the southern one. Although the origins were different, there developed similar shapes because environmental conditions and ecological niches were similar.

Could parallel evolution occur on planets that are similar but very distant from one another? It is possible. One day, perhaps—hundreds of years from now—our newly started research in interstellar communications will bear fruit and our descendants will see on a television screen a manlike being that is not the creation of a Hollywood studio but a real alien living on a distant planet.

Without giving up hope for such a wonderful event, let us return to what we can learn from our current research. First of all, we have found an answer to the question we raised when we discussed the last days of the earth. Most likely, the end of life on our planet will not mean the end of life in the universe. By that time intelligent terrestrials may have migrated to other planets like ours, or be living in space colonies, or be traveling among the stars. But even if this did not happen, even if terrestrial life should end with our planet without leaving an issue, it would not mean the total extinction of life. Every year life probably sprouts on 7 planets in our Galaxy and on 70 billion worlds throughout the universe. Thus when the earthlings die out there may be planets where life is just starting and planets where it is already highly advanced, having begun earlier, though later than on the earth.

Finally, there is one more point to discuss that concerns both the past and the future. In order to calculate the number of planets where life may be sprouting every year, we began by assuming that it has been happening since the birth of the Galaxy. Then we raised some questions. Are we sure that life processes began when the very first stars were born? Are we sure that at that time there formed also planets similar to the current ones? Perhaps, as we assumed on second thought, all this happened much later, possibly shortly before the appearance of life on earth. If it is so, there would have been a lifeless age in the universe, after which life began, evolved, and spread at an accelerating pace.

There is only one way to find out if life originated at the same time as matter or began after a long period of preparation—we must explore the universe from the most remote epochs: before, long before, the sun and the earth were born. On the other hand, this is also the way to find the solution of the main problem: How and why was the universe born?

7   TOWARD THE ORIGIN OF EVERYTHING

THE ANCIENT STARS

All we have seen thus far happened in the past 5 billion years, is happening now, or will happen in the near future. When we started off 5 billion years ago we knew that the universe had already existed for at least 10 billion years. We have not yet explored those 10 billion years. What happened in that long period of time? One might think of events similar to those we have already seen: formation and destruction of stars by the same processes and at the same pace that we have discovered for the current stars; origin and evolution of planets much like those in the solar system; development of living beings similar to the past, present, and future inhabitants of the earth. But we cannot assume that the present can be extended by projecting backward, more or less cyclically, what is happening today, and we cannot be entirely sure that in the remotest past things occurred in the same way as today, even though by "today" we mean the last 5 billion years.

Indeed, we know that it cannot be so because even in the "recent" past terrestrial life and the earth as a whole have been continually changing in a noncyclical way. Furthermore, if it is true that the universe shrinks as we go back in time, we must reach a moment when it was very small. Obviously such a universe could not contain either stars or galaxies but only matter at a very high density—and before then, not even matter. If we want to learn what the universe was like at that time, what existed before and after that moment during that very long period of more than 10 billion years, we must set off on another journey in time, long before the birth of the sun and the earth, and head toward the crucial moment when everything began.

THE GLOBULAR CLUSTERS

The first step in this direction was taken, unintentionally, through the study of the globular clusters, and it was a decisive step. These star clusters are composed of hundreds of thousands of stars distributed with nearly spherical symmetry around the center of the cluster (see figure 4). Some are found in the galactic disk, where the sun also is located; others in the nearly spherical region known as the galactic halo.[43]

Until the mid-1950s, the H-R diagrams (magnitude-color arrays) of globular clusters seemed unaccountably different from those of open clusters (figure 71). Both exhibited a conspicuous group of red giants, but in the globular clusters there were no blue giants. Thus the brightest stars in globular clusters were not blue, as in open clusters or in the sun's neighborhood, but red. Moreover, there was a sequence of stars nearly parallel to the horizontal axis of the diagram that for this reason was called a "horizontal branch." Last, and strangest thing of all, there did not seem to be a main sequence.

In 1956, however, A. R. Sandage used the 5-meter telescope on Mount Palomar to contruct a new H-R diagram for the globular cluster M 3 reaching stars as faint as the 21st visual magnitude, that is, 63 times fainter than the cluster stars previously observed with less powerful telescopes. Plotting the diagram, he discovered that there was a main sequence, but it started from stars much fainter than those normally found in the H-R diagrams of open clusters (figure 72). Furthermore, the main sequence veered off uninterruptedly into the red-giant branch, as observed in some open clusters.

This diagram is characteristic of all gobular clusters, and two common features—a low turnoff point and a main sequence continuous with the red-giant branch—made them correspond to the oldest open clusters. One would think that by appropriately superimposing the main sequence of a globular cluster on the main sequence shown in figure 20, one could immediately find the age of the cluster from its turnoff point. In fact there is a small complication, which, ironically, led to a discovery that revealed a past of the universe no one had ever suspected.

Let us consider this small complication. Since the first spectroscopic observations it was known that the stars of most globular clusters consist almost exclusively of hydrogen and helium. If you recall, the chemical composition of a star's interior varies continuously because hydrogen turns into helium, helium into carbon and oxygen, and sometimes, depending on the mass of the star, into heavier elements. Such variations do not affect the composition of the outer layers (those we observe) because the outward spread of the elements produced in the interior is extremely slow. Consequently, from the spectroscopic analysis of a star we do not find its average composition at the time of observation, but rather the composition of the star at the time it was born. In other words, the chemical composition we observe is that of the material from which it was formed. The sun, almost all the stars in

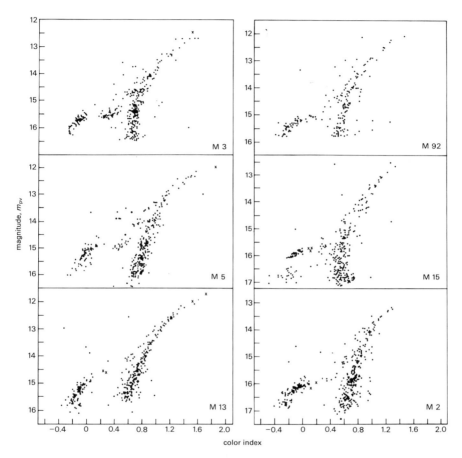

Figure 71   H-R diagrams for six globular clusters published by H. Arp in 1955. Note the absence of the main sequence in all of them. Clusters are denoted by the numbers assigned to them in Messier's catalog. [From *Astronomical Journal* 61 (1955):330]

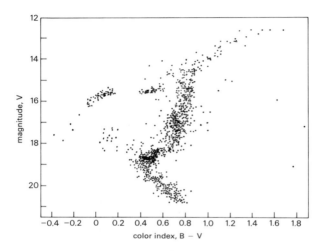

Figure 72   The H-R diagram for the globular clustere M 3 prepared by John-
son and Sandage in 1956. By observing faint cluster stars with the Mount
Palomar 5-meter telescope, they were the first to find the main sequence in the
H-R diagram of a globular cluster. Since m_{pv} and V are about the same, this
diagram is directly comparable with the diagram for M 3 shown in figure 71.
[From *Astrophys. J*. 124 (1956):379, © The University of Chicago Press]

our neighborhood, and the stars in open clusters, unlike globular-
cluster stars, show not only hydrogen and helium but also most of the
other elements, though in much smaller amounts. Thus the former
were formed from a material rich in metals,[44] while the latter were
formed from a material consisting almost exclusively of hydrogen
and helium. Because of their different initial composition, by the Vogt-
Russell theorem, globular-cluster stars fall on a main sequence that is
different from that of open-cluster stars. Thus, to determine the age of
a globular cluster we cannot simply superimpose its main sequence on
that of the diagram in figure 20; we must plot another diagram that
takes into account the difference in initial composition (mainly hydrogen and helium).

The difference in initial composition is the key point. Theory shows
that after the helium flash, metal-poor stars move rapidly downward
and to the left-hand side of the diagram; then, more slowly, they again
swell into red giants traveling from left to right, at first horizontally and
then steeply upward. Figure 73 shows the evolutionary path calcu-

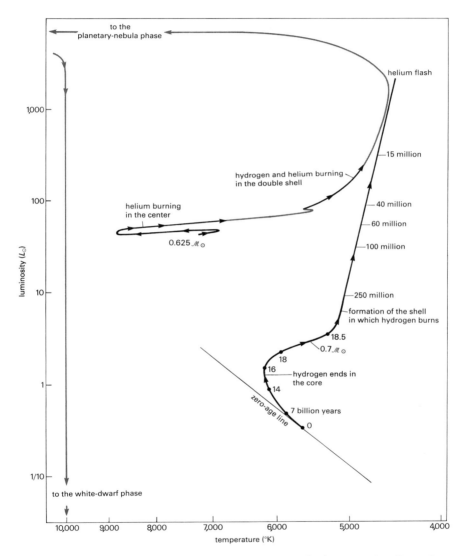

Figure 73   Evolutionary paths of a metal-poor star of $0.7M_\odot$ starting from the zero-age line. Note the horizontal branch. [Courtesy the Astronomical Society of the Pacific]

lated by I. Iben for a star of $0.7 M_\odot$ with an initial composition of 70% hydrogen, 30% helium, and 1‰ metals, which is the composition of stars in globular clusters. The structure and evolution of such a star is discussed and illustrated in detail in appendix 1. As figure 73 shows, metal deficiency causes a star to go through the evolutionary stages corresponding to the horizontal branch. On the other hand, subsequent theoretical studies have clearly shown that metal-rich stars cannot go through those same stages. Hence the H-R diagrams of open clusters, made of metal-rich stars, cannot show the horizontal branch; at most, it is reduced to a fragment protruding from the red-giant branch and often indistinguishable from it.

This explains the strange H-R diagrams for globular clusters shown in figure 71. The apparent lack of a main sequence is due to the advanced age of the clusters, which lowers the turnoff point, and to their great distance from us, which makes it possible for only the largest telescopes in the world to see the brightest stars that make up the faint remaining piece of main sequence. The presence of the horizontal branch is a consequence of the different evolutionary history of metal-poor stars. Finally, the abundance of red giants is due to the fact that they are the most numerous among the brightest stars in globular clusters, which are the stars that can be seen even with less powerful telescopes.

At this point it is interesting to determine when the open and globular clusters were formed, that is, to find their ages. This has been done, case by case, with the method, already mentioned, of the position of the turnoff point in the H-R diagram. Although the method is conceptually simple, there are some practical difficulties, some of which we have already mentioned. In the first place, one must take into account the fact that globular and open clusters do not have the same zero-age main sequence due to the difference in their initial composition. Furthermore, the globular clusters are so far away from us that it is difficult to observe enough faint stars to obtain a significant portion of the main sequence and the turnoff point; this is possible only with the world's largest telescopes and for a part of the known globular clusters. We rarely have this problem with the open clusters, but here there is another difficulty. The main sequences of all the open clusters are calibrated to that of the nearby Hyades cluster, and therefore any difference in the determination of the distance to the Hyades affects the distance and age of any of the others. Thus it is not surprising to find that different authors furnish different results. However, the latest estimates are generally the most reliable.

The extensive research on star clusters of both types has produced a surprising result. Most of the open clusters are young or very young, generally ranging in age from a few million to a few hundred million years. This means that many of them were born when the first human beings were already living on the earth. Some, like the cluster in Orion, are still forming. Until a few years ago the oldest known were M 67 and NGC 188. According to the results published by B. A. Twarog in 1977, they are 3.5 and 4.3 billion years old, respectively—a venerable age, but still less than that of the sun and the earth. The globular clusters, instead, are very old, ranging in age from 13 to 18 billion years. Hence they would have formed soon after the birth of the universe.

A comparison between the ages of globular and open clusters leads to a discorcenting conclusion: Between the formation of the last globular clusters and the formation of the first open clusters there was a long period of time in which no stars were born. In 1979 Twarog and two of his collaborators found that the oldest known open cluster is not NGC 188 but Melotte 66, whose age would be between 6 and 7 billion years. Even considering this extreme case, there would remain a period of at least 6–8 billion years in which no stars were born in our Galaxy. Why? Why is it that millions of stars formed rapidly right after the universe began and then no more were born for a period of at least half the time elapsed from the beginning of the universe to the present? What happened in the meanwhile? On the other hand, how can we be sure that no more stars were born during that period? Perhaps there formed only single stars, or clusters so poor that they soon broke apart. In either case it would be very difficult to track down these vagabonds of space and calculate their ages.

There is only one way to solve the problem, and that is to trace the evolution of the Galaxy from its birth to the present.

THE EVOLUTION OF THE GALAXY

STELLAR POPULATIONS
Globular and open clusters do not comprise all the stars of the Galaxy. As anyone can easily see with the naked eye from our observation post

inside the Galaxy, there are myriads of scattered stars. In the 1940s, however, it was found that these stars are roughly of two types: some are like the stars in open clusters, and some like the stars in globular clusters. Two distinguishing characteristics are age and chemical composition. The third is space distribution, which will give us the first clue to the evolution of the Galaxy.

Most of the stars we see with the naked eye are distributed in space in such a way as to form a disk 100,000 light-years across and about 6,000 light-years thick. The sun and the earth orbiting about it are also inside this disk, which we see as a luminous band across the sky, or Milky Way. Open clusters and stellar associations, as well as the dust and gas from which new stars are still forming, are found only in this disk. Nearly all of the globular clusters, along with a number of scattered stars, are distributed instead here and there, at great distances, more or less symmetrically with respect to the center of the Galaxy, in a roughly spherical region 80,000 light-years across. The stars and globular clusters that form this sort of sphere, called halo, are more concentrated in the central part than at the periphery (figure 74). A very important characteristic of the galactic halo is that it does not contain any gas or dust from which new stars can form. This means that its stars formed in the distant past. The stellar population of the disk was called Population I because, being closer to us, it was known before we discovered the halo population, which was called Population II.

A comparison of the two populations revealed yet another difference. Population-I stars travel in space at low velocities with respect to the sun, while Population-II stars move at high velocities. This is essentially due to the fact that Population-I stars revolve around the center of the Galaxy at nearly the same speed as the sun, always remaining within the disk, whereas Population-II stars revolve about the galactic center in highly elliptical orbits oriented in all directions. When these stars are in the outer regions of the halo they move very slowly, but as they approach the galactic center they move faster and faster (by Kepler's second law). While they cross the disk they appear to move at great speed, partly because they are actually moving faster in this portion of their orbits, and partly because they move in different directions from that of the sun. A practical example may help to clarify the concept of low- and high-velocity stars. If you drive at 90 km/hr and overtake

Figure 74   Schematic drawing of the Galaxy showing the disk, halo, open clusters, and globular clusters. Galactic objects (stars, clusters, etc.) in Population I are in the disk, while those in Population II are in the halo.

a truck moving at 80 km/hr, the truck seems to move slowly, but the same truck would appear to move much faster if, still traveling at 80 km/hr, we saw it coming from a ramp more or less inclined to the highway's direction or, worse yet, if it cut across the highway.

Taking into account all the characteristics of the stars of Populations I and II, at the end of 1962 O. J. Eggen, D. Lynden-Bell, and A. R. Sandage were able to formulate a theory for the formation of the Galaxy as it is now that since then has been widely confirmed.

FORMATION AND EVOLUTION OF THE GALAXY

Eggen, Lynden-Bell, and Sandage found that there is a close correlation between the chemical composition of stars and the types of orbit they travel in the Galaxy. Metal-poor stars always travel highly elliptical orbits, while metal-rich stars like the sun travel circular or nearly circular orbits. Furthermore, the lower the metal content, the higher is the component of the velocity perpendicular to the Galaxy's equatorial plane, and the smaller the angular momentum of the orbit. The three scientists concluded that metal-poor stars with highly elliptical orbits formed very early in the life of the Galaxy. According to their theory, updated with more recent data, this is how it happened.

As a consequence of the early evolution of the universe, by 15 billion years ago a large number of gas clouds had formed, of various sizes, in slow rotation, and composed almost exclusively of hydrogen and helium. From these clouds would be born the galaxies, including our own (figure 75). At first our protogalaxy must have been a vast cloud about 130,000 light-years across, in equilibrium, since the pull of gravity toward the center was balanced by the outward pressure of the gas, as we have seen in stars. Then, by the well-known process of aggregation, stars began to form in the regions of the protogalaxy where gas density was higher, and the first globular clusters were born. This caused gas pressure to fall, and the remaining gas began to collapse toward the center of the Galaxy. During the in-fall of the gas, lasting about 200 million years, there formed more globular clusters and stars of the halo population.

The collapse of the protogalaxy along the equatorial plane, that is, the plane passing through the galactic center and perpendicular to the rotation axis, was soon slowed down and halted by the centrifugal effects of rotation, but the collapse along the rotation axis and inter-

Figure 75   Diagram showing how a galaxy like ours could have formed. (a) A vast cloud of primordial gas, consisting essentially of hydrogen and helium and in slow rotation, cools down. (b) Halo stars, globular clusters, and a central nucleus of stars are formed; the cloud of primordial gas is still there, but its density keeps on decreasing. (c) The residual gas falls toward the center, but in the regions around the plane perpendicular to the rotation axis the fall of the gas is first slowed and then halted by the centrifugal force; meanwhile, the first massive stars have reached the end of their evolution and have exploded, enriching with heavy elements the primordial gas falling toward the equatorial plane. (d) The in-falling gas settles around the equatorial plane, forming a disk from which a new stellar population will be born. (e) The first massive stars formed in the disk have quickly evolved and exploded as supernovae, starting chains of explosions that form the spiral arms and an increasing number of stars, including stars of small mass. (f) The galaxy has reached the stage where its structure is similar to the present structure of our own galaxy; the less massive old stars of the halo continue to exist, isolated or in globular clusters, while in the disk there develops a population of celestial bodies (stars, planets, etc.) similar to those we first discovered in our solar system.

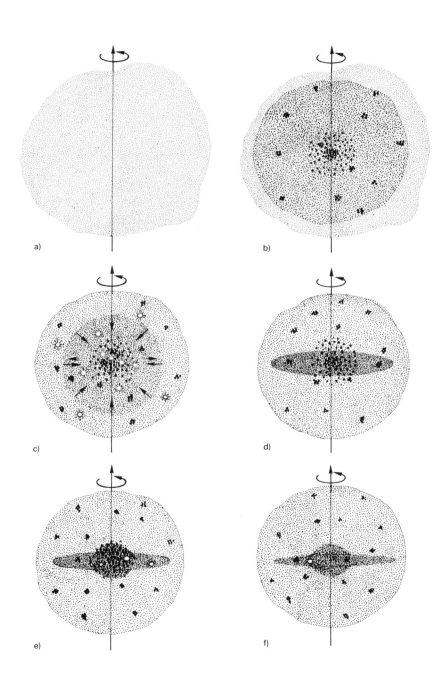

mediate directions did not stop until all the gas of the halo had been packed into the present disk. While new stars continued to form from the gas massed in the disk, in the halo were left only the first stars and globular clusters, still traveling the vast elliptical orbits of the gas from which they had formed.

In the 200 million years during which the collapse lasted, the heaviest of the first stars had completed their evolutionary cycle, which for stars of 5 solar masses lasts less than 100 million years and for stars more than 9 times the sun's mass lasts less than 25 million years. During that time they had built elements heavier than hydrogen and helium and had exploded as supernovae; in the course of the explosions they had also built the elements heavier than iron. Thus the newly forming stars were not composed of hydrogen and helium alone, but of new material enriched with heavy elements. Recalling that the spectroscopic analysis of a star's light yields the chemical composition of its outer layers, which still consist of the original material, this explains why the second-generation stars of Population I are much richer in heavy elements than the first-born stars of Population II.

According to this model, the globular clusters that formed at the end of the collapse of the gas halo should already contain an appreciable amount of heavy elements, and in fact they do. We know several globular clusters that are part of the disk. They are the last to be born, from enriched material, and unlike their brothers in the halo, they are rich in metals. Since the halo population mixes with the disk population in the region they have in common, it is possible to see, next to each other, a globular cluster that is metal-rich and one that is metal-poor. A case in point are the globular clusters M 69 and M 54 (figure 76). The former is rich in heavy elements and actually belongs to the disk; the latter, metal-poor, is a halo cluster that is traveling its elliptical orbit and appears close to the other by chance and only temporarily.

At first it seemed natural to think that metal enrichment was a continuous, gradual process. This hypothesis seemed to be confirmed by the fact that not all stars fit exactly into Population I or Population II. In fact, these two are only extreme cases, the most common and evident. In 1957, during a meeting organized by the Pontifical Academy of Sciences, five distinct stellar populations were defined; beginning with the oldest, they are halo Population II, intermediate Population II, disk Population, intermediate Population I, and extreme Population I.

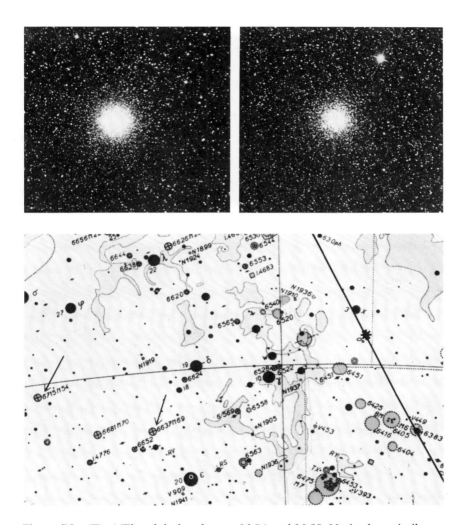

Figure 76   (Top) The globular clusters M 54 and M 69. Notice how similar they are despite the difference in their chemical composition. (Bottom) A section of the celestial atlas by A. Bečvář shows that the two clusters are close to each other and to the galactic equator (diagonal line at right; the asterisk marks the galactic center), even though M 54 belongs to the halo and is only passing through this region. [Top, courtesy L. Rosino; bottom, from A. Bečvář, *Atlas Coeli,* Prague: Academia]

However, according to the research published by the Soviet astronomers V. A. Marsakov and A. A. Sukhov between 1976 and 1978, the evolution of the Galaxy was not a continuous process. By grouping the globular clusters according to the abundance of metals, these scientists concluded that they fall into three distinct groups. Subsequently, they found that scattered stars also could be grouped in the same way, not only with regard to metal content, but also on the basis of certain orbital characteristics. All these findings indicate that the entire population of the Galaxy can be divided into three groups, whose characteristics are in agreement with the described evolutionary process but show that the process was not continuous. According to Marsakov and Sukhov, during the collapse of the protogalaxy there were three moments of explosive activity in which metals enriched the interstellar gas, and two intermediate periods in which this did not occur. Stars formed in these periods but not in the others. And this conclusion, as mentioned, applies not only to the clusters but also to isolated stars. Thus it was a general phenomenon and it always involved the entire Galaxy. This model is supported by the results obtained by Demarque and McClure, already mentioned, and by the evidence that B. Tinsley has been gathering for some years. Starting in 1973, moreover, various astronomers, pursuing different lines of research, have found that something of the kind seems to have happened in other galaxies as well.

On the other hand, star formation is a process that is still going on, and by studying it on a spatial and temporal scale vaster than the formation of small groups of stars we can better understand the past, in addition to learning how the Galaxy is evolving at present, that is, in the past few billion years. In fact, by extending our observations to the entire Galaxy we can trace stellar evolution to ages close to the beginning of the universe.

THE SPIRAL STRUCTURE

In our first journey through time we witnessed the formation of stars from the contraction of large clouds of gas and dust. Since this material is initially scattered over vast expanses and extremely tenuous, it must first be compressed into clouds of the right consistency. Only then can the clouds break up into the fragments that will contract into dark globules with the necessary characteristics for stars to be born. First of all, in other words, the gas and dust must be sufficiently compressed for gravity to take hold. This can occur in various ways, two of which are particularly important because they may even determine the structure of the Galaxy, and not only that.

As you know, what we have been referring to as the galactic disk is not really a disk, at least today, but a spiral structure. The spiral, which is beautifully evident in external galaxies (figure 3), has been reconstructed for our galaxy as well, though with some difficulty since we are inside of it.[45] Theoretical studies have tried to explain how the spiral arms were formed and how they are maintained or continually reconstituted.

According to a theory formulated by C. C. Lin and F. H. Shu in the early 1960s, the spiral features are caused by a density wave passing through galactic matter at a velocity lower than the velocity at which the Galaxy's components, such as stars and dust, revolve about its center. It is a very simple phenomenon—we all cause it every morning by stirring a cup of coffee. Attrition causes the liquid near the cup's edge to slow down and lag behind the liquid near the center, just as differential rotation does with the Galaxy. Thus beautiful spires form in the swirling coffee, which stand out even better when beads of cream ride on them. As one of Lin's former students, W. W. Roberts, subsequently demonstrated, a shock wave traveling ahead of the density wave may compress the interstellar matter to 5–10 times its original density, triggering gravitational contraction in the denser clouds and setting in motion the process of star formation. When these stars burst into incandescence, their light, along with the light of the nebulae they excite or illuminate, makes the spiral arms visible, much like the beads of cream in the coffee.

At first it seemed that the density-wave mechanism could account for the formation of only two spiral arms, but recent theoretical studies indicate that it could produce up to four spiral arms. On the other hand, the spiral structure can be explained by another theory that, contrary to Lin's, starts from the formation of stars and arrives at the formation of spires.

According to Lin, it is a density wave that compresses the gas to the density needed to trigger the formation of stars. Back in 1953, however, E. J. Öpik had argued that the gas could be compressed by the shock wave generated by a supernova explosion. In 1977 W. Herbst and G. E. Assousa announced that they had found a region where this

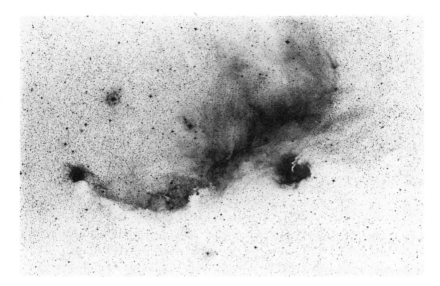

Figure 77   A photograph of the region R1 in Canis Major showing some of the very young stars observed along the outer edge of the nebular ring; this is where the interstellar material has been compressed by a shock wave and gives rise to new stars. [Photograph taken by W. Herbst of Wesleyan University at the Cerro Tololo Interamerican Observatory]

process is at work, namely, the association C Ma R1, which contains dozens of very young stars. A great many of these stars fall above the zero-age main sequence in the H-R diagram and are no older than 300,000 years. The extraordinary thing about these infant stars is that they are located at the edge of a vast nebular ring 220 light-years in diameter that also contains neutral hydrogen and is expanding at the rate of 13 km/sec (figure 77). According to Herbst and Assousa, a supernova that exploded 500,000 years ago sent a shock wave that swept up the surrounding material and compressed it into the still-expanding ring; 200,000 years after the explosion this material began to generate stars that presumably are still being born.

This case is not unique because others were subsequently found, such as Mon R1 and another in the association Cep OB3. Furthermore, what we are seeing is only the beginning. Among the new stars formed from the material swept up and compressed by the shock wave generated by the supernova, there must be at least one massive enough to

evolve in a few tens of millions of years and itself explode as a supernova. This, in turn, will generate a shock wave that will sweep up more
interstellar matter, at least in the regions that had not been "cleaned
out" by the previous explosion. Thus the phenomenon will be repeated
again and again. Actually, since several massive stars will form along
the great arc of compressed gas, all ending as supernovae, the phenomenon will be amplified in a chain reaction that will stop only when all
the interstellar matter in the region has been swept up, compressed,
and transformed into stars.

Studies carried out by H. Gerola and P. E. Seiden since 1978 show
that the structure of galaxies and even the existence of certain types
of galaxies, such as the dwarf ones, can be explained by this kind of
chain reactions. They assumed that the galactic disk is initially uniform
(spiralless), that it rotates according to the law of differential rotation,
as all galaxies do, and that supernovae explode in it, which must surely
happen as a result of stellar evolution. They fed these data into a computer with the instruction that the explosions should be entirely random and, naturally, that the laws governing the mechanics of fluids
and the propagation of waves should be taken into account. This computer simulation of galactic evolution produced spiral galaxies altogether similar to the galaxies observed by the hundreds in the universe
(figure 78).

Thus we have another mechanism that can account for the formation
of stars and the spiral structure of galaxies. Perhaps it coexists with the
mechanism proposed by Lin; it may prevail in certain types of galaxies
and explain anomalies observed in the spiral structure of many others.

THE OTHER SIDE OF THE COIN

The globular clusters were formed in the remote past, at the beginning of the universe, while the open clusters date from an epoch much
closer to our own. The general mechanisms of formation of clusters,
particularly globular clusters, are not yet clear in detail. Knowledge
of these processes would certainly be useful to our understanding of
galactic evolution, but it is not really essential. What is important is
that clusters do exist; it has been demonstrated that they did not form
from scattered stars that came together by chance, and we have even
glimpsed some mechanisms by which open clusters and associations
may form.

Figure 78   (Left) A computer simulation of a galaxy. (Right) The spiral galaxy M 101 photographed at Mount Palomar. The computer simulation, due to H. Gerola and P. E. Seiden, is based on the theory of chains of supernova explosions. Notice the similarity between the model and the real galaxy. [Left, from *Astrophys. J.* 233 (1979):65, © The University of Chicago Press; right, courtesy Mount Palomar Observatory]

On the other hand, we know quite a bit more about the future of clusters. Sooner or later, they are all destined to break apart, scattering their stars far and wide. The globular clusters, rich in stars and hence very massive, can last longer because the gravitational attraction that holds them together is very strong. In addition, they reside prevalently in the halo, where they are isolated and less subject to the disruptive forces of large external masses, which tend to pull the stars away. By contrast, the open clusters, containing at most a few thousand stars, cannot last as long. Since they are located inside the disk, they also are disrupted by the differential rotation of the Galaxy, which causes the cluster stars closer to the galactic center to move faster than the stars on the opposite side of the cluster.

Taking into account all the forces that can disrupt a cluster, we estimate that the life span of a globular cluster is of the order of magnitude of 10 billion years, while an open cluster will last some billion years and an association will be dissolved in a few tens of millions of years.

This means that a number of globular clusters, at least the poorest, many open clusters, and an enormous number of associations have dissolved since the birth of our Galaxy. According to V. G. Surdin, as many as 2,000 globular clusters may have disbanded for a total of 300 million solar masses; but of course many of their stars would still be in existence, scattered throughout the galactic halo. As we have seen, some of the oldest open clusters, such as M 67, NGC 188, and Melotte 66, still exist as such 5–6 billion years after they formed. Almost certainly these clusters were originally so rich in stars that their strong gravitational attraction has been able to hold them together for a long time.

In 1978, on the other hand, V. M. Danilov demonstrated theoretically that in the last stages of its life a cluster evolves in a nonclassical way. Depending on its conditions, a cluster may expand, accelerating the dispersion of its stars, or it may stabilize or even contract. The fate of a cluster may be influenced by the presence of galactic gas, dust clouds, and binary systems. Thus it is not clear that richer clusters will necessarily last longer. In the same year another Soviet astronomer, Yu. N. Efremov, announced that he had discovered several stellar complexes that in a not too distant past must have occupied extended regions where thousands upon thousands of stars were born and that today are dispersing. Up to 1978 he had identified 35 such groups with an average diameter of 2,000 light-years and an age of some tens of millions of years. Based on the study of the Cepheid variables found in each group, Efremov estimated that the process of star formation in such complexes lasts from 20 to 50 million years.

The dynamic picture of the evolution of the Galaxy from its origin to the present begins to appear fairly clear, at least in its broad lines. But all we have seen would not have led to the present world if the Galaxy had not undergone a less conspicuous but much more radical process of transformation, namely, a chemical evolution.

THE FORMATION OF THE PRESENT WORLD

As we have learned from the globular clusters and from the evolution of the Galaxy, when the first stars were born in the vast gas cloud that was the protogalaxy, there were essentially two elements at hand—

hydrogen and helium. We can assume that in the other galaxies forming in the early universe things were going the same way and that the material available to the galaxies in order to form had that same chemical composition. Had the universe continued to be made almost exclusively of hydrogen and helium it would not be what it is today, and none of the things familiar to us would exist. But our world is not made of hydrogen and helium, and we must search the past for the cause of this extraordinary transformation. At this point it is not hard to find. By now we know how the stars evolve and what has transformed, and continues to transform, our Galaxy and a great many of the others. It all seems clear and yet absolutely amazing.

The first stars, those we still see today in the globular clusters, were formed from hydrogen, helium, and a few atoms of other light elements. In their interiors hydrogen burned into helium, helium into carbon and oxygen. The more massive stars burned carbon and oxygen into elements as heavy as silicon and iron. Many stars became supernovae and in the course of the explosion generated elements heavier than iron, which they injected into space, among the clouds of primordial hydrogen, along with all the other elements they had built before the explosion. But it was not only the supernovae that enriched with new, heavy elements the hydrogen clouds left in interstellar space. The red giants also build heavy elements in their shells. But while in supernova explosions element building occurs through the capture of rapid neutrons, in the red giants it occurs through the capture of slow neutrons; the two processes are denoted, respectively, r (rapid) and s (slow). Furthermore, not only the way they build part of the elements is different, but also the way they inject them into space. Whereas the supernovae enrich the interstellar gas in a violent outburst, the red giants do it quietly but continuously through the process of mass loss. There are other processes whereby stellar matter can be dispersed, some of which we have seen earlier, such as the explosive burning of carbon and oxygen cores, and others we shall not go into. Thus if we exclude the stars of very small mass that never reach the main sequence and expire even before they are born, all the stars in the course of their lives build and disperse into space heavier and heavier elements that enrich and change the clouds of hydrogen and helium. As we clearly see and as demonstrated by facts, this enrichment of the interstellar matter is irreversible. We do not know any natural process whereby the elements built from hydrogen can be transformed back into hydrogen and helium.

Let us continue our survey of cosmic evolution. New stars formed from the new enriched material, and at a certain point during this long process of development different celestial bodies also began to form, such as the planets and all the other cold and solid bodies—in sum, all the material we know in the present universe, which gave rise to everything around us. We need only reach out a hand and touch what is nearby: a book, a pair of glasses, a stone, a gold ring, a china cup, an ashtray, the ashes and cigarette stubs in it. None of this would exist if it had not been built over time by the stars. But why reach out a hand? We can as well sit still because all we have to do is think. There would not be the air we breathe or a good deal of the matter in our bodies. We ourselves would not exist.

When we discussed where and how life is born we found strong evidence that it begins with prebiotic molecules. Then we wondered whether the conditions and the elements necessary for life to appear existed at the time the first stars formed in our Galaxy. Now we know that they did not. With hydrogen, helium, and trace amounts of deuterium and lithium there could not form prebiotic molecules, which are based on carbon, oxygen, and other elements—and there could not exist solid, complex planets like our own where life could develop.

After the formation of the first stars there was a period in which life could not start anywhere, not even in space, for lack of elements, and then there was another period in which the elements were being built but as yet there were no places where life could take hold. We do not know how long these periods lasted or how much they overlapped. All we know is that by about 5 billion years ago both periods were over.

We had also wondered what happened in those 6–10 billion years in which everything seemed to have come to a stop. Now we have finally seen it. We know what happened starting from about 15 billion years ago when the massive stars in the first globular clusters began to explode and to enrich with heavy elements the gas of the halo that was still collapsing toward the equatorial plane of the Galaxy. What happened was something fundamental that would cause a qualitative change in the fabric of the universe—through their lives the first-generation stars of Population II were paving the way for the modern world and our life. By 5 or 6 billion years ago this new world was taking shape, and in the last few million years life has developed to the point of leading us to this culminating moment of reasoned contemplation of our past.

TIME

THE FAILURE OF THE CYCLES

What we have just learned about the past of the universe erases even the last trace of a cosmic cycle, namely the transformation of interstellar matter into stars and planets and then back again into gas and dust from which new stars are formed. Not even this cycle has existed forever. Tracing the history of the universe from the remotest ages, we find that it is a one-way street and a discontinuous process subject to changes that sometimes have been slow and gradual, sometimes quick and dramatic. We have discovered that despite appearances, events not only never recur cyclically but are irreversible. This causes an asymmetry in time.

We can better understand this property by comparing time with space. In space, we can decide to move one way or the exact opposite way. In time this is not possible—we cannot choose to move toward the future or toward the past; we must always move toward the future. If you think about it, everyday life, monotonous and repetitive as it may seem, is also made of unique, irretrievable moments. Like the life of the universe, it flows in one direction in a way that is determined by all that happened before.

We are about to hear a concert. The event has been prepared by scores of people who have devoted a good deal of their lives to the study of music. They have gathered here for a purpose, the same purpose that brings us here, in this hall, with these lights, with these particular acoustics, studied and arranged by technicians and a stage manager. After long years of training, the voice soloists have studied, assimilated, and rehearsed the musical score we are about to hear; they have discussed it with the other musicians, and now they are going to perform it with their own voices and in their own interpretation. The performers have all come from different cities for this event. Tomorrow they will be in some other places, wherever their engagements call them, but now they are here with us and for us. The conductor mounts the podium and leads us to the sublime moment of the creation of a work of art that will be unique and, by the very nature of music, will vanish as it is created. Hundreds of people, some visible and some invisible, contribute to this creative moment: the composer, who is long dead; the performers; the men who built the hall and the instruments; and we ourselves, because by connecting the notes reaching our ears we give substance to the musical masterpiece that, as long as it remains in writing, has only a potential value, that is, does not exist in its true essence—sound.

We may hear this work again, but it will not be the same, not even if we heard it the very next evening, in the same hall, and with the same performers. Even if everything else remained the same, we would have changed—if for no other reason because of the effect of the previous performance.

This is true for everything, for every moment of pleasure or pain, joy or sadness. Prepared by an intricate set of circumstances that are lost in the past, each moment ripens, unfolds, and is gone forever. It can never happen again because all the circumstances that have brought it about can never be the same. None of us can experience the same thing twice in the same way, just as the earth could never pass twice through the same point in space even if its orbit remained the same.

Returning to the cosmic view, we realize that we have learned something else. Each of the unique events of our lives has been in preparation since the universe began. The splendid concert we have just heard would have never been performed if the universe had retained its original composition of hydrogen and helium and if, between the beginning of time and today, there had not been the whole sequence of events that by now we know almost in its entirety.

And another thing. Just as all that happened in the past has shaped our world and our lives, we ourselves, in our lives, help to shape the future of the universe, even if only within the small expanses of space and time allotted to us. We are transient specks of dust, a consequence of the past of the universe, but at the same time, along with countless other specks of dust, we are the shapers of its future.

COSMIC TIME

Time is asymmetrical; time is acyclical; finally, time is relative to each observer. We have not yet encountered this property of time, but the theory of relativity has demonstrated it, and we must accept this result even though we cannot stop to discuss it. In all our excursions into the past we have used the time marked by our clocks and calendars, but this was all right only so long as we did not wander too far from the earth or too far back in time. In this last journey, however, we continued to use our earth time while approaching ages very close to the beginning of the universe. That is not proper. According to relativity, time is different for a being on another galaxy because our two galaxies are in motion with respect to each other. We cannot accept a reconstruction of the universe based on his time any more than he can accept one based on our time. It is not a matter of pride, but a real, practical difficulty in that we cannot describe the history of the universe as a whole using different times for different places. Instead, we must find a time whereby each instant is the same everywhere. More precisely, we must define what is meant by simultaneity.

To arrive at such a definition we shall start from the cosmological principle, which states that "on a large scale the structure and properties of the universe are the same everywhere." Suppose now we place observers at various points in the universe and that the motion of each observer, with respect to the others, is such that it respects the cosmological principle. This can occur only if the distances between the observers remain the same, or if they increase or diminish uniformly. In this definition the ideal observer is one who is at rest with respect to the galaxy in which he is located. Now we can introduce a time to which we can refer the evolution of the entire universe. It is the time of the ideal observers, each motionless in his own galaxy. In other words, it is the time marked, at each point, by a clock at rest with respect to the matter at that point.

This time is called *cosmic time*. With it, we can redefine operationally the concept of simultaneity and thus describe the transformation of the universe as a whole.

An application of cosmic time may serve to explain it better. We know that the universe is expanding and that therefore its average density must progressively diminish. By the cosmological principle, in the same instant the average density must be the same everywhere in the universe. Thus we shall say that two or more events are simultaneous when, the moment they occur, two or more observers, each at rest with respect to his own galaxy, measure the same average density in the universe.

It follows that it is no longer proper to express time in the astronomical units that, though the first to be used, have lately proved not to be entirely reliable. Naturally we can continue to measure time in years, minutes, and seconds, but these intervals must be defined, as indeed has been done for years in everyday life, by natural mechanisms (the frequency of a spectral line, radioactive decay, etc.) that, as we shall see, are governed by laws we can trust to be valid at all times and at all points of the universe.

Cosmic time is not an absolute time, which would be incompatible with the theory of general relativity, but it is a universal time provided the cosmological principle is valid. If it were not valid—for example, if we found that half the galaxies were receding and half were stationary—then cosmic time would no longer make sense. But this is not the case, and by using this temporal reference we shall go back as close as possible to the instant the universe began, and then forward into the most distant future, or, more precisely, into some of the possible futures that at present we are able to conceive.

We have seen that the universe is not a static entity, immutable in all its vastness and variety; it is not cyclical—that is, though changing, it keeps on repeating itself; finally, it is not stationary, in the sense that while its individual components change (people, planets, galaxies, etc.) it remains always the same as a whole.

The universe does change. The universe has a history. And it is the true history of the world. Any other history, such as that of mankind, or the earth and everything that happened on it, or a galaxy, or a man, is just a small episode in it. When you read a history book, the life of a common individual seems brief and insignificant compared with all the events and personalities that have been handed down to us. Yet the entire recorded history of humankind is only a hundred times longer than the life of an individual, while the history of the universe is already more than a million times longer than that of humankind. Furthermore, both the history of a man and that of mankind up to now have unfolded in a very small and insignificant place when compared to the vastness of today's universe.

By now we have several proofs that the universe in the past was different from the present one. First of all, as you will recall, there has been a change in chemical composition—the older stars were formed almost exclusively from hydrogen and helium, while the younger ones are rich in metals. A second proof is given by counts of extragalactic radio sources. These sources are often associated with quasars and giant galaxies, both of which are intrinsically more luminous than common galaxies. They are so bright that they can be seen many billions of light-years away, while the galaxies, with the same instruments, are normally observed to distances of up to about 3 billion light-years. Counts of these radio sources show that they become more numerous the farther away we look, but at a certain point their number drops off abruptly. Since looking billions of light-years in the distance means looking billions of years in the past, this discovery tells us that earlier in the life of the universe there were more radio sources, and that earlier yet there were none at all. To put it another way, there was a time when there were no radio sources in the universe; then there were many; and then, as we approach the present, they become fewer and fewer, perhaps because they have turned into something else.

Another proof that the universe has changed is the fact that it is expanding, which today is accepted by virtually all the astronomers and cosmologists. At present we see the universe made up of widely separated galaxies, but if, by reversing their motion, we make them retrace their steps, as we go back in time they will move closer and closer to one another until, inevitably, we shall reach a moment when all the matter contained in the galaxies was bunched together. Thus at least at that time the universe must have been quite different from what it is now. And it may have been different earlier on, when the space containing all matter and energy was even smaller.

We shall now explore these first moments of the universe in an attempt to understand what happened right after the beginning and, if possible, at that very moment.

THE FIREBALL

In our second journey in time we went back to when the galaxies were born, and we discovered that the material that formed their first stars consisted essentially of hydrogen and helium. Now we intend to discover the reason for that chemical composition, what was there before, and what happened between the Big Bang, when the universe began, and the epoch in which the first galaxies formed. Finally, we intend to measure how much time elapsed between the initial moment and that epoch. To do this we must go further and further back in time. As we shall see this journey will not last very long. But these will be the most difficult years to explore because we are much less helped by observations and because in the course of a few years the universe went through states thoroughly different from one another and from the current one.

INCONCEIVABLE TEMPERATURES

In the distant past the universe must have been very much hotter than it is now. All bodies warm up when compressed and cool down as they expand. This law also applies to the universe as a whole, and we have a formula that gives its temperature with the passage of cosmic time. When we apply it, we find that in epochs close to the initial instant, when all matter was compressed into a sort of sphere of enormous density, the temperature was fantastically high: thousands, hundreds of thousands, millions, billions, billions of billions of degrees. The closer we get to the moment of the explosion, the higher the temperature.

Cosmologists have named that early universe the *fireball*—a figure of speech that is very evocative but wholly inadequate to describe reality. Nevertheless, this figure of speech helps to "translate" reality into terms that we can comprehend from our own experience. When we read a book translated from a language we do not know we may lose the shades of meaning of the original and even misunderstand some passages, but the overall meaning is clear and we understand the things said in it even though we cannot savor how they were said. The same happens when we read the first pages of the history of the universe. We shall envisage events according to today's experience and images, but the true sense of what happened will not be lost.

Suppose we stand looking at the sky while time rushes back toward the beginning of the universe. At first the night sky alternates so quickly with the day sky that we do not see it—our eyes retain only the bright blue of day. Then the sky darkens rapidly, the soil vanishes from under our feet, and we remain standing in space. We have reached the time when the sun and the earth have not yet been born. Now the sky is dark and dotted with thousands of stars. The background remains dark while the stars change, move, turn on, and turn off at a dizzying pace until we reach a time when they all disappear, including the myriads of stars that formed the Milky Way. Only a few nebulous dots are left, scattered here and there. We are about 15 billion years in the past; the Galaxy has not yet collapsed and the stellar population of the disk does not exist. There are only the faint stars scattered in the halo, which we do not see, and many globular clusters, of which only the nearest and richest are visible as nebulous dots.

The sky is still dark—darker than before because the background stars, though invisible individually, contributed a slight, diffuse luminosity. In this epoch the galaxies are much closer together and hence to us. But like our Galaxy they are just forming and cannot be seen. Another step backward and even the galaxies are gone; now there is only primordial gas, invisible, diffuse in space. We see nothing at all. The sky is dark and empty. But not for long. As we continue to go back in time, the sky begins to glow a deep dark red—it is the primeval fireball. Since we, too, are inside of it, we do not see it in any one part of the sky; rather, it is the whole sky around us that has lit up. Gradually, the sky turns a brighter and brighter red, becomes blazing hot, increases in splendor as it turns orange then yellow; by now unbelievably luminous, it tends to change color again when, almost abruptly, everything disappears. The fireball has vanished. Now we are immersed in an impenetrable fog, and had we been able to make this trip in reality rather than in our fantasy, we would have become fog as well. Why?

To understand all this we must retrace the very first phases in the evolution of the universe. We can do it only theoretically, taking as experimental data those on the structure of matter and on the forces that govern the universe as we know them today, on the assumption that they have never changed and that they are valid everywhere. This reconstruction of the first ages of the world should not be considered either complete or definitive. However, some of the events we shall be able to reconstruct had consequences that can be traced and often have been actually found in today's universe. This proves that our reconstruction is sound in its broad lines and perhaps in many details as well.

THE 3°K RADIATION

One of the fundamental proofs that the universe is expanding and that it began in a mighty explosion was given by the demonstration that the fireball really existed.

As G. Gamow pointed out as far back as 1946, if we assume that the universe is expanding, it follows that initially all matter was compressed into a very small space. Hence it must have been very hot because any gas heats up when compressed. It was by following this reasoning that in our rush through the past we saw the universe get hotter and hotter.

In reality, of course, the sequence of events occurred the other way around. The instant the universe exploded the temperature was highest and then it began to fall rapidly. The temperature (as we shall see better later on) was not that of any body we know, such as a star or an aggregate of stars; nor was it confined to any particular region. It was the temperature of the entire universe, that is, the temperature of the universe at each of its points. As the universe expanded, the temperature dropped because it became diluted within a volume that was ever increasing, but it did not reach absolute zero, which is unreachable. Today, therefore, all points of space should be at the same low temperature, the last residue of the enormous temperature of the fireball, and each point of the universe should radiate according to this temperature. Thus at any point in the universe—for instance, on the earth—we should measure a background radiation that, corresponding everywhere to the same temperature, should have the same properties in all directions. To use the proper scientific term, it should be isotropic.

This isotropic background radiation was detected in 1965 by A. Penzias and R. W. Wilson. Although the discovery was expected and hoped for by many cosmologists, it happened by chance. Penzias and Wilson were radio astronomers, but at that time they were working at Bell Labs, at Holmdel, NJ, perfecting an antenna for satellite communications. While testing the antenna for background noise, they detected a faint but persistent static that could not be eliminated by pointing the antenna in different directions. Convinced that the noise was due to a defect in the instrument, they rechecked it but to no avail —the signal was still there, still coming from all parts of space.

While Penzias and Wilson were investigating the strange noise, R. Dicke and his collaborators at Princeton were building a very sensitive radio telescope in the hope of detecting what Penzias and Wilson had just found: an isotropic radio emission, that is, an emission coming from all parts of the sky with the same characteristics. Like Gamow, Dicke and his group believed that if the universe had once had a very high temperature, no matter how long ago, it should retain a trace of it that one could detect. As soon as Penzias and Wilson became acquainted with the research of the Princeton group, they knew they had found what Dicke and his collaborators had predicted and were seeking.

Subsequent research showed that this background radiation was not exactly the same in all directions, but the discrepancies were soon explained. They are due to our motions, that is, the motion of the earth around the sun and that of the solar system within the Galaxy. Furthermore, in 1976 it was found that the Galaxy itself, besides participating in the general expansion of the universe, is moving in the direction of the constellation Leo at 600 km/sec. Later on we found that the entire group of galaxies in our neighborhood, within a radius of about 10 million light-years, is moving in the same direction.

In sum, the radiation discovered by Penzias and Wilson fills the entire universe uniformly, and the earth is immersed in it much like your body in the warm water of the bathtub. If you move your hand in the water and feel the liquid flow over your skin, this is not due to the motion of the water but to the motion of your hand. By measuring the speed with which the liquid flows over your skin in effect you measure the speed with which your hand is moving in the water. The same happened with the background radiation. At first, when distortions were observed, they were found to be due to the various motions of the earth, already well known and measured by other methods. A further asymmetry revealed the motion of the group of galaxies to which our own belongs. Thus the fundamental property of the background radiation, isotropy, was conclusively established.

The background radiation has enabled us to make a most important measurement, namely, that of the current temperature of the universe.

All bodies emit energy in the form of electromagnetic waves, which, depending on their wavelengths, are detected in different ways. If the wavelength is rather long, 1 mm (millimeter) or longer, the emitted electromagnetic waves are detected as radio waves. Between 1 mm and 1 m radio waves are known as microwaves. Between 1 mm and 0.8 μm (micrometer) they are perceived as infrared radiation. Between 0.8 μm and 0.3 μm they are seen as light, whose color depends on the wavelength. Finally, at wavelengths shorter than 0.3 μm they appear as ultraviolet rays, x-rays, or γ-rays. The energy a body emits at each wavelength depends on the body's nature, surface area, and temperature. With regard to the body's nature, everything else being equal, the body that can emit the greatest amount of energy is the so-called black body, which is also the body that can absorb the most energy. You can prove this fact to your satisfaction by walking out into the hot sun, first with a black suit, and then with a white one. A perfectly black body does not exist in nature, but it can be approximated quite well in the laboratory. With regard to the surface area, we can assume it is equal to 1 by always referring to a unit of surface area—say, a square centimeter or a square meter.

Given the above, if a black body is heated to a certain temperature, it will emit a certain amount of energy as electromagnetic waves that do not all have the same wavelength but span an interval of wavelengths. If we plot the energy emitted at each wavelength we obtain a curve whose shape depends on the body's temperature. Figure 79 shows the types of curves that would be obtained, known as Planck's curves. Notice that the peak of emission is different for each curve: The higher the temperature, the higher is the peak and the more it shifts toward the shorter wavelengths. This is why bodies at high temperatures, above 3,000°, for example, appear luminous and of various colors (the

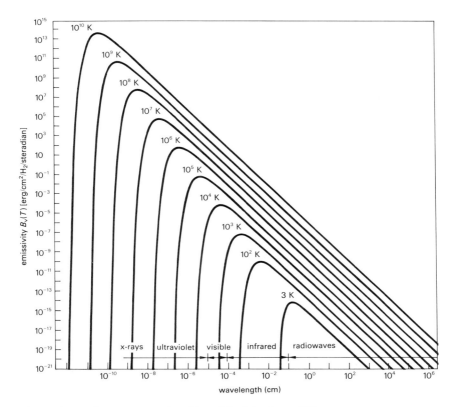

Figure 79   The energy emitted by a black body at different temperatures. Planck's curves indicate the relative energies radiated in different wavelengths for each temperature (°K). Observe that with increasing temperature the peak of emission shifts toward shorter wavelengths, i.e., from the red to the blue. [From K. R. Lang, *Astrophysical Formulae*, 2nd ed., Springer Verlag, 1980, p. 22]

bluer, the hotter they are), while cooler bodies can be detected only with instruments sensitive to the infrared or radio waves. Thus, to determine the temperature of the body, even a distant one, we need only plot its energy curve, that is, measure the amount of energy it emits at different wavelengths, and then find the black body curve that best fits the observed curve. It is in this manner that we measure the temperature of the stars. This same method has been used to measure the present temperature of the universe.

When Penzias and Wilson discovered the background radiation their instrument was tuned to the wavelength of 7.35 cm. Hence they measured only one point in the energy curve. Six months later Dicke's group made another measurement at the wavelength of 3.2 cm. Subsequently enough measurements were made that the energy curve could be plotted (figure 80), and it was found that the current temperature of the universe is 2.7°K.

In 1979, however, D. P. Woody and P. L. Richards made new measurements from space with more sophisticated techniques. They found that the spectrum of the background radiation does not exactly fit a black body spectrum. More precisely, the shape of the curve is the same as that of a black body at the temperature of 2.79°K, but the amount of radiation they detected is 1.27 times greater than a black body would emit at that temperature. I. E. Segal has suggested an explanation of this anomaly with a nonconventional cosmological theory, but the matter is not yet settled. At any rate we should not lose sight of the fundamental discovery—the fireball. By now the fire is very faint, almost expired, but it is still burning, and it will always burn even though it will continue to grow fainter and fainter, demonstrating to us that the universe must have expanded and that once it was enormously hotter.

From today's temperature we can determine the temperature of the universe in times past and find how the shape of the corresponding black body curves have changed over time (figure 81). Observe that at present the universe radiates exclusively in radio waves. Earlier it still radiated in the radio range of the spectrum but at shorter wavelengths. Earlier yet the universe radiated mostly in the far infrared, then in the infrared, and then in visible light. If you recall, as it was becoming ever bluer and more luminous, like a metal brought to incandescent heat, almost abruptly all light failed. Yet the universe was certainly very hot and growing hotter as we approached the moment of the great explo-

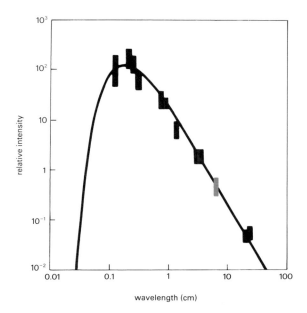

wavelength (cm)

Figure 80   The intensity of the background radiation measured at different wavelengths by different workers. Note that the measurements fall on a black body curve. The length of each bar indicates the margin of error in the measurement. The bar approximately above the wavelength of 10 cm corresponds to the measurement by Penzias and Wilson. [From *Le Scienze* (June 1980)]

sion. The reason it went off must be sought in the structure of the newly born universe, shaped by the matter, forces, radiation, and physical laws that still exist today. Hence we must discuss these things, if briefly, in order to obtain the key to the ultimate observatory from which we shall observe our world in the first moments of its existence.

THE STRUCTURE OF THE UNIVERSE

As everyone knows, all the matter in the universe, from our bodies to the stars and the intergalactic gas, is made of atoms. Each atom, in turn, consists of smaller particles (electrons, protons, neutrons, and so on). The various elements have different chemical properties, depending on the number of subatomic particles and the way they are grouped together. The early universe was too hot for atoms to exist;

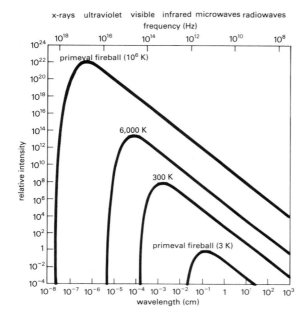

Figure 81   Black body curves indicating the temperature of the universe at different times. The top curve refers to the primordial fireball soon after the Big Bang. The bottom curve corresponds to the background radiation observed today. The other curves refer to intermediate ages. [From *Le Scienze* (October 1968)]

there could only be particles, which in turn must have formed from energy. Although we cannot investigate directly the particles and energy of that time, we believe we know fairly well the particles that now make up the atoms and the forces that govern all of matter. We shall deduce from these the structure and transformations of the early universe.

THE PARTICLES
When, at the beginning of this century, scientists began to penetrate the atom, believed to be indivisible until the last decade of the nineteenth century, they thought it was composed essentially of three types of particles: electrons, protons, and neutrons. The electron carries a negative electric charge and is very light; the proton carries a positive

electric charge numerically equal to that of the electron, but is 1,836 times heavier than the electron; the neutron weighs as much as the proton but has no electric charge.

In the first part of this book, when we were looking for an atomic clock, we briefly discussed how these three particles combine to form the elements. This simple model of the atom enabled us to build all the elements (neutral or ionized) and their isotopes by varying the proportions of only three units. Since these particles were considered the basic building blocks of matter they were termed "elementary particles."

Since 1945, however, the number of known particles has grown rapidly, and now we have dozens of them. By the 1960s there already were so many that it was felt necessary to group them in families of similar characteristics. Today all subatomic particles are distributed in two classes: the more numerous group comprises the heavier particles, or *hadrons,* which include protons and neutrons; the second is the small group of light particles, or *leptons,* which include the electron and the neutrino. The latter is a particle with no electric charge and with a mass that is zero or very small. To complete the picture, we should add a third group, which consists of a single component, the *photon;* this is a particle with no mass or electric charge but endowed with energy. Rather than a particle, the photon actually should be considered a packet of electromagnetic waves that, depending on its wavelength, appears as a γ-ray photon, an x-ray photon, a light photon, etc. But the photon does behave like a particle in the sense that when it interacts with something—our eyes, for example—it does so as a unit. We shall never see just half of the packet of waves that constitute the photon; either we see all of it or none of it, just as if it were a particle.

In the early 1960s, as he was grouping the many known particles according to their affinities or similarities, the American physicist M. Gell-Mann went beyond this by asking himself whether all the particles hitherto believed to be indivisible might not in fact be composed of yet smaller particles of a few types that, by combining in different ways and proportions, would explain the different kinds of leptons and above all the large number of known hadrons. Gell-Mann postulated three particles with new characteristics, including a fractional electric charge, and called them *quarks.*[46]

Even though later on the quarks grew in number, Gell-Mann's sub-

atomic model remains very convincing. There is a problem, however—
as yet no quark has been isolated experimentally. It is not a simple
thing to do because the forces that bind quarks into elementary particles are very strong. (Of course, if Gell-Mann is right, elementary particles would lose any claim to the term "elementary," which by rights
should go to the quarks.) Thus to isolate quarks it is necessary to apply
very high energies to the particles. One day this may be possible, but at
the moment their existence is still in question, particularly because theoretical studies of the early universe argue against it.

ANTIMATTER

Our brief excursion into the world of particles is not yet over. Everything around us and the universe itself might be constructed, so to
speak, the other way around. As predicted theoretically by P. A. M.
Dirac in 1928, and as later demonstrated experimentally, for each
known particle (hadron or lepton) there is another one identical to it in
mass but opposite to it in electric and magnetic properties. These particles are called *antiparticles*. The first of them was discovered in 1932 by
C. D. Anderson. It had the same properties as an electron, but carried
a positive electric charge; it was named *positron*, or positive electron.
The antiproton was discovered in 1955, and afterward, one by one,
scientists found the particles "opposite" to all the known ones.

Antiparticles could form atoms entirely similar to those we know, but
they would have a negative nuclear charge surrounded and neutralized
by positive electrons. In turn, such atoms could form aggregates of
matter with the same characteristics as the matter we know and are
made of. In sum, there could exist bodies made of antimatter, including very large bodies—for example, a galaxy. If there were such a
galaxy, we would have no way of knowing that it was made of antimatter. We know a galaxy only from the light we receive from it, and the
photons would be the same in either case because there is no anti-
photon. In other words, the photons cannot tell us whether a body
consists of matter or antimatter. On the other hand, a galaxy made of
antimatter is not likely to exist because when particles and antiparticles
meet they annihilate each other, transforming 80% of their masses into
energy. This world of ours, composed of matter, makes life practically
impossible for antimatter, which can be built only in our laboratories,
in the form of very short-lived antiparticles.

THE FORCES

Our world would not exist if all matter, from the smallest particles to the largest aggregates, such as planets, stars, and galaxies, were not bound and governed by certain forces. To the best of our knowledge, there are four forces that act in the universe and have largely contributed to determine its structure and evolution. The physicists call them by a more descriptive and appropriate term—interactions.

Strong Interaction   As we have seen, atomic nuclei can be formed of numerous particles, some of which, like the protons, carry an electric charge of the same magnitude and sign. While two electric charges of opposite signs attract each other, two electric charges of the same sign repel each other. How is it possible, then, for a number of protons, all positively charged, to stay together in the tiny atomic nucleus instead of flying off and disrupting the nucleus? Something must keep them together. This "something" is the strong nuclear interaction, the strongest force that acts in nature, 137 times stronger than the electromagnetic interaction that would tend to separate the protons. It is this force that binds the hadrons together into a stable atomic nucleus.

If this force is so powerful, one might ask, why does it not pull together all the existing hadrons and reduce the universe to one huge atomic nucleus? The reason is simply that it has a very limited sphere of action; it does not act beyond a sphere whose diameter is $1/10^{15} = 1/1,000,000,000,000,000$ of a meter. In practice, this force is effective only within the nucleus of the atom. It is powerful enough to overcome all the other forces and impart the greatest strength to the building blocks of the universe (atomic nuclei), but so limited in extent that it cannot tear away blocks from nature's constructions and destroy the marvelous architecture of the universe by reducing everything to a single huge block.

Weak Interaction   It is the force that acts on leptons. It has the same sphere of action as the strong interaction (in practice, that is, it acts only within the atomic nucleus), but it is 10^{13}, 10,000,000,000,000, times weaker.

Electromagnetic Interaction   This is the familiar force that governs all electromagnetic phenomena, from the repulsion or attraction of electric charges to the heat and light we receive from the sun. We put it to work in countless ways—for example, in the motors of our appliances, our radios, and our television sets, to mention a few. As noted, it is 137 times weaker than the strong interaction, but it is enormously more intense than the weak interaction. Unlike both, its sphere of action is not confined to the atom but extends to infinity; this force decreases as the square of the distance, but though becoming weaker, it never ceases. This seems to be contradicted by the fact that in nature there are countless charges and the action of one appears to be annulled (though it is not) where another begins to prevail. In effect, they interact with one another. To complicate matters even more, such interactions may be attractive or repulsive, depending on the sign of the charges.

Gravitational Interaction   Gravity is the force that gives us weight, making life so tiringly human; it is the force that holds the planets in their orbits, moves the stars in the galaxies, and the galaxies in the universe. Moreover, it is the only force capable of counteracting the expansion of the universe. Yet it is the weakest of the four known forces, 10^{38}, or 100,000,000,000,000,000,000,000,000,000,000,000,000, times weaker than the strong interaction. Of course, to earth-bound creatures it seems very strong, so strong, in fact, that for millennia it hid from us the action of the other three forces. It is not difficult to see why. First of all, like the electromagnetic interaction it has an infinite radius of action. Furthermore, unlike the latter, it is always positive; that is, it always acts as an attraction. It is true that the force of gravity decreases rapidly with distance, but it is also true that the more massive a body, the stronger is the force. These are the reasons that have made it so "popular" since time immemorial. Although this force is the best known, the other three forces play just as important a role in the workings of the universe—if only one of the four had been missing or had a different value, the universe would not be what it is today.

LAWS AND PRINCIPLES
The various components of the universe, from particles to galaxies, and the forces that govern their interactions have been discovered by means of a vast number of observations logically connected and rationally interpreted. In this manner we have formulated laws that in the

form we express them exist only in the context of man's perception of the universe, but are essential to the advancement of science. Their great value lies in the fact that they permit us to understand the world beyond the limits of the experiments from which they were deduced.

To acquire their power, laws must be based on principles, that is, statements suggested by experience and logic or imposed by necessity. Take, for example, the law of universal gravitation. As a "law" it does not exist in nature because, as general relativity has demonstrated, gravitation is merely a property of space-time. Based on Galileo's experiments and on his own logical interpretation of certain phenomena, Newton was able to formulate the law in its general form with a simple expression. Calling the law "universal" was a statement of principle dictated by necessity. In order to extend our knowledge of the universe, the law had to be applicable beyond the narrow confines of the solar system, and its universal validity had to be posited. On the other hand, logic and the Copernican conception that there are no privileged points in the universe made such a generalization quite reasonable.

As science advances, laws and principles may be updated, introduced, or discarded. Two opposing principles may also exist at the same time, supported by different scientists and leading to opposing theories. The choice between the two may then be dictated by observational evidence favorable to one or contrary to the other, or simply by necessity, which may come down to a matter of opinion.

A classic example is the choice between the cosmological principle and the perfect cosmological principle. The first, as you recall, states that "on a large scale the structure and properties of the universe are the same everywhere." The second states that "on a large scale the structure and properties of the universe are the same everywhere and at all times." The statement that all properties, including the mean density, are constant in time implies a continuous creation of matter and leads to a steady-state universe, that is, a universe that has always existed and as a whole (though expanding) does not change. Twenty years of research have demonstrated that this is not so, and in our journey in time we have seen why. But even before observational evidence argued against it, many scientists were loath to accept the perfect cosmological principle because it conflicted with another principle— conservation of matter. This, too, is based on experience but above all on an essential requirement of our thought: Something in the universe must be permanent in character, or we cannot possibly arrive at a rational conception of nature. If we give up such a principle, any research would make about as much sense as a game of cards played with the knowledge that one of the players is cheating.

ON THE QUESTION OF UNIVERSAL VALIDITY

All we have learned about the structure of matter, the forces that govern it, and the laws that guide us in our quest for knowledge, preventing us from getting lost in the maze of the universe, has been discovered or tested in our laboratories or, at most, in that portion of the solar system that has been explored by man and his probes. We have no direct proof that our results are valid everywhere else and at all times, past and future. Yet, ever since we began to discuss the stars in our Galaxy we have assumed that they were all valid at distances of a few light-years as well as of thousands of light-years. For example, when we study the motions of the stars in a binary system we assume that they are ruled by the same law of gravitation we have tested on the earth. From that same law, and by assuming that Kepler's laws of planetary motions are applicable to the motions of the stars, we have calculated the mass of our Galaxy.

The results of our studies paint such a coherent picture of the world of stars that we feel entitled to maintain that the laws found on the earth are also valid for a distant globular cluster and that the matter in the interior of its stars is in the physical conditions predicted by our theoretical physics and tested in our laboratories. But how can we be sure that this is true for the most distant galaxies and even for the matter and forces at play in the universe before stars and galaxies were born, in epochs so remote from us and so close to the extraordinary moment at which the universe began? In that instant everything must have been entirely different, even if in a relatively short time the universe became quite similar to the current one in its general outline.

Although the universal validity of our laws was once accepted as a principle, in recent years we have tried to prove it. Extensive research has been done on the spectral lines emitted by laboratory light sources and by distant objects. From the ratio of the wavelengths or the intensities of two or more spectral lines one can obtain information concerning the structure of matter and some physical constants. Thus, by making these measurements for terrestrial light sources and for a distant galaxy, depending on whether the results coincide or not, the universality of the result obtained on the earth can be proved or disproved.

In 1980, from spectroscopic observations of four quasars located at different distances and hence corresponding to different ages, A. D. Tubbs and A. M. Wolfe found that the strong interaction and the electromagnetic interaction were the same as today as far back as the time when the universe was 5% of its present age, that is, when the universe was about 700 million years old. Thus it seems that we can indeed apply today's results to the universe of the past.

Having said this, we can finally try to reconstruct what happened before the time the galaxies began to form.

THE BIRTH OF THE UNIVERSE

If the universe is expanding, the distance between two of its points, say two galaxies, varies with time. Thus as we go back into the past its total mass becomes concentrated into a smaller and smaller volume and the average density of matter becomes higher and higher; actually we have a formula that allows us to calculate it moment by moment. From observations we deduce that its current value is about 3×10^{-31} g/cm^3 (grams per cubic centimeter). As we approach the moment of the explosion the average density increases at such a vertiginous pace that the universe becomes superdense. The value of the density at the exact instant of the origin cannot be known because we cannot reach that moment even theoretically.

According to the theory of general relativity, the closest we can get to the origin is 10^{-43} second, or 0.0000000000000000000000000000000-000000000001 second, after the Big Bang. The density then must have been about 10^{94} g/cm^3 (figure 82). That is a very impressive value but, after all, not unexpected. What is strange is something else. Just like matter, the radiation that fills the universe has its own density. At present it is about 10^{-33} g/cm^3, almost all of it being the diluted original radiation. Going back into the past, with another suitable formula we find that as we get close to the beginning, the density of radiation increases faster than that of matter. Thus there was a moment at which radiation and matter had the same density, and before then, an age during which the universe consisted much more of radiation than mat-

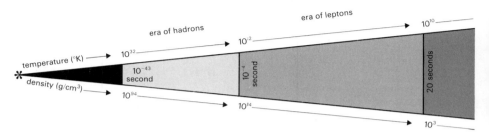

Figure 82   The four eras of the universe. [Adapted from *Astronomy* (August 1979)]

ter. Until a few hundred-thousandths of a second after the explosion the distinction between matter and radiation did not exist at all. Earlier, what was extremely dense was neither matter nor radiation, but only energy—the energy that in time would become all the matter, all the radiation, and all the energy that exist today in the entire universe.

As we have seen, the temperature also was much higher in the remote past. By applying the formula that gives the value of the temperature as a function of time, we find that 10^{-43} second after the explosion it was 10^{32}°K, or 100,000 billion billion billion degrees. Through the mathematical formulas that give density and temperature for increasing cosmic time, and a third formula that yields the energy of radiation as a function of temperature, we can attempt a reasonable reconstruction of what happened when the universe was dilating at a speed that is inconceivable, even if we compare it with the speed of such violent explosions as a hydrogen bomb or a supernova.

As noted earlier, our reconstruction can describe what actually happened to a good approximation on the assumption that the physical laws we know today hold even for the first moments in the life of the universe. Astronomical observations cannot take us back that far, and we cannot observe what occurred in the crucial seconds that determined the whole future life of the universe. However, some of the consequences predicted by theory have been verified by observations of today's universe. The existence of the 3°K radiation is a case in point.

We should also note that the infinitesimal fraction of a second

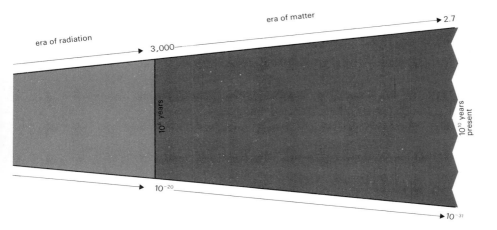

(10^{-43}) in which it is impossible to know what the universe was like is a theoretical limit. In effect there are uncertainties and conflicting views for a much longer period, a millionth of a second. Why is that first fraction of a second so impenetrable even to theory? There is a reason. The large-scale universe, with its immense distances, is ruled by universal gravitation and explained today by the theory of general relativity. The structure of matter, which is made of atoms, of very tiny particles, is explained by quantum physics. In that first fraction of a second the laws governing these two opposites—the infinitely large and the infinitely small—had to be valid, not only at the same time, like now, but in a certain sense for the same object. In fact, the whole universe was perhaps reduced to a point where we must apply the laws for the immensely large, and where the particles described by the physics of the infinitesimal were about to be born. Applying both theories as space shrinks schematically to a point and time falls to zero is an awesome undertaking, fraught with difficulties that appear insurmountable. Obviously, the difficulties are only in our minds because, ignoring relativity and quantum physics, the universe was born.

After its birth the universe changed radically more than once, going through different eras. The first two were very brief, lasting fractions of a second and a few seconds, respectively, while the most recent era has been going on for at least 15 billion years. But it is not duration that counts. The important thing is what happened in those eras, brief as they may have been. Once again, then, let us set off on a journey in time, a journey that, though the shortest of them all, will be quite eventful, a journey in which we must not forget that all we are about to see, particularly at the beginning, is still a matter of controversy within the scientific community. In order not to be confused by the many conflicting views, we shall go by those that are more generally accepted or more promising.

THE FOUR ERAS OF THE UNIVERSE

We do not know what there was at the beginning or how the universe was born, but we do know that all we see today condensed in celestial bodies or diluted in space either did not exist in those first moments or must have been enormously concentrated. Of course, there had to be the four interactions that today rule the world, and with such a restricted sphere of action the forces had to be at very close quarters, so to speak—perhaps unified in a single force. The first to be singled out was the force of gravity, which separated from the others shortly after the universe had emerged from that unfathomable instant lasting 10^{-43} second.

The Era of Hadrons   In this era both the temperature (10^{32}°K) and the density (10^{94} g/cm^3) are so high as to be simply inconceivable. At such a high density, all possible particles form and annihilate in equal measure. Quarks and leptons are born, all with their respective antiparticles, but it is above all the heaviest hadrons that form. They are born from the energy packed into enormous amounts of γ-ray photons so energetic that they materialize essentially as baryons and antibaryons, the heaviest particles and antiparticles among hadrons. No sooner have baryons and antibaryons formed than they come into contact, again because of the high density, and annihilate each other, turning into energy.

Statistically, baryons and antibaryons should have formed in equal numbers. Then they *all* would have been transformed into energy, and the world of matter in which we live and of which we are part would never have existed. But it did not happen that way. For each billion pairs there was a baryon left over that survived as such. The tiny fraction of heavy particles that survived constitutes the near totality of the matter (which is not little) distributed today in the entire universe.

Due to the lightning expansion of space, in a very short time the temperature drops to $10^{13°}$K, or 10,000 billion degrees. In the interval of time between one-millionth and one ten-thousandth of a second after the Big Bang, the weak interaction, the electromagnetic interaction, and the strong interaction separate in turn. Henceforth the four forces and their realms will be quite distinct. At this point the γ-ray photons no longer pack enough energy to form baryons and antibaryons; their production stops, and the surviving baryons turn into neutrons and protons. A ten-thousandth of a second has elapsed from the Big Bang. The density has dropped to 10^{14} g/cm^3 and the temperature to $10^{12°}$K, or 1,000 billion degrees, too low for even the lightest of the hadrons, the π mesons, to form. The formation of hadrons comes to a halt. Their era is over. It has lasted less than a ten-thousandth of a second, and a fundamental event for the formation of the current world already has occurred.

The Era of Leptons   As the temperature drops the lighter particles form, and with so few hadrons left, it is the leptons' turn to dominate the universe. Hordes of electrons and positrons are born that for the most part annihilate each other, forming photons that expand and will continue to expand throughout the universe. A type of leptons, the neutrinos, have less and less to do with matter, decouple, and begin an independent existence, eventually filling the universe like the background radiation.

But the most interesting story is that of the remaining hadrons. At the beginning of this second era the number of neutrons and protons is very nearly the same. Then the neutrons decay in large numbers, freeing protons and electrons. Thus the protons become more and more numerous. Meanwhile, many neutrons have combined with protons to form nuclei of deuterium (heavy hydrogen), each consisting of one neutron and one proton. Now pairs of deuterium nuclei begin to combine, forming helium nuclei, each of which, as you know, consists of two protons and two neutrons. It has been calculated that by this process, which soon became more efficient, hydrogen and helium formed in the percentages of 80% and 20%, respectively.

Now we finally understand why the ancient stars in globular clusters are composed almost exclusively of hydrogen and helium. We also have solved another problem that we had not yet mentioned, namely, why the current percentages of helium and hydrogen in the universe, stars, and nebulae are, by mass, 25–30% and 75–70%, respectively. We have seen that helium forms from hydrogen in the interior of stars. But if helium had formed only in this way, today it would be just 3–5% of the hydrogen. Instead, it is a little more than what we find in the ancient stars. And the marked difference between what it should be and what it is corresponds precisely to the quantity that, according to theory, turns out to have formed from the fusion of deuterium nuclei during the era of leptons.

Thus we have another confirmation of the theory. Also, we have ascertained that the oldest stars consist almost exclusively of hydrogen and helium because those were the only elements abundantly present in the material from which the first stars were formed, and that material was so constituted because it had formed that way during the Big Bang.

The Era of Radiation   The era of leptons lasts about 20 seconds. It is between the end of this era and the first 100 seconds of the third era that helium nuclei are formed. Meanwhile, in those first 20 seconds, the universe has expanded enough for the temperature to drop to 1 billion degrees and the density to a "mere" 1,000 g/cm^3. As a result, the formation of matter from energy comes to a stop. Protons, neutrons, and electrons, along with all the other particles, are afloat in a sea of photons a billion times more numerous. Photons are packets of electromagnetic waves, i.e., radiation. Hence radiation dominates this new era, which lasts 300,000 years.

In the first 15 minutes, matter consists of 75% protons and 25% helium nuclei, besides an almost negligible number of lithium nuclei, composed of 3 protons and 3 or 4 neutrons. After 10,000 years the temperature drops below 10,000 degrees; the existing nuclei begin to capture electrons, forming the first neutral atoms of hydrogen, helium, and lithium. Meanwhile, the universe continues to expand and photons continue to fill it, but since the universe is growing larger, they become more and more diluted. After 300,000 years the density of radiation, heretofore higher, becomes the same as the density of matter. Just for a moment. Henceforth, though the photons will remain a billion times more numerous than particles, the density of radiation in the universe becomes increasingly lower than that of matter. Naturally they both continue to decline because the universe becomes larger and larger and no new matter is born, but the density of matter decreases more slowly than that of radiation.

At this point the period of the fireball is over. The evolution of matter and that of radiation now proceed independently, and with the decoupling of matter and radiation the universe at last appears. Up to this time, the universe, terribly hot, full of radiation and hence light, was in reality completely dark. Radiation and light had indeed filled the universe to the point of giving the name to one of its eras, but trapped by matter like water in a sponge, they could not be perceived. With the decoupling, light springs free and the universe can be seen: first white-hot, and then, as the temperature decreases, yellow, orange, red, deepening to infrared, until, billions of years later, it will emit only radio waves of ever longer wavelength.

Thus 300,000 years after its birth the universe lights up, only to darken again in a few million years—or so it would have appeared to our eyes if we had been there to watch. Meanwhile, matter, which is everywhere diffuse, begins to condense; soon it will concentrate, forming galaxies filled with stars that will become, and will remain for billions of years, the new source of light.

The era of radiation is finished. Now begins the fourth era—that of matter—which is still going on.

The Era of Matter   In the view of some cosmologists the moment at which matter and radiation decoupled and the universe became transparent coincides with the moment at which the density of radiation fell below the density of matter, and they hold that this occurred when the universe was 1 million years old (see figure 82). By this time the density has dropped to 10^{-20} g/cm^3 and the temperature to 3,000 degrees. The sky still emanates a reddish light that can cross undisturbed the clouds of hydrogen and helium, which are still everywhere diffuse.

The first galaxies are about to form, and all we have observed in our previous journeys in time is about to take place. But how and why did protogalaxies form from the primeval gas, protogalaxies within which stars would then be born that would form the galaxies we see today in all their variety? We have seen the universe evolve to the point of building the first light elements. How did this featureless universe evolve into the universe we live in? This transition is one of the most obscure passages in the history of the world. It has been rightly defined as the interface between two sciences, cosmology and astrophysics. These difficulties, however, will not prevent us from exploring it, at least to the extent of extracting what seems most certain from the tangle of current knowledge.

THE BIRTH OF THE GALAXIES

On 14 and 15 February 1979 a workshop was held on the formation and early evolution of galaxies organized by W. H. McCrea and M. J. Rees for the British Royal Society. As McCrea himself later remarked, "When the discussion started almost certainly no one present would have claimed to know how any galaxy was formed; neither would anyone who was present at the close." Nevertheless, that meeting was very important for galactic studies because it revealed how little we know about the galaxies, what we do know for a fact, and what can be done to learn more.

One thing is certain from the start: The galaxies have not always existed, but began to form in a certain evolutionary phase of the universe. Twice we came close to that moment: first, when we went back to the time of the formation of the first stars of Population II; and again a short while ago when we came forward in time to follow the evolution of the universe from the beginning, from a time when not only were there no galaxies but there was not even the matter to form them. We also know that by whatever process the galaxies formed, it must have been a very efficient one because it used up nearly all of the matter in the universe. Let us consider some evidence for this statement.

Today we know both galaxies that are isolated and galaxies that are massed in clusters. The first x-ray observations of the sky with the Uhuru satellite revealed that the intergalactic spaces within a cluster, once believed to be empty, actually contained gas that might have been a residue of the primordial gas from which the galaxies of the cluster had formed. Moreover, there was also the space between clusters to explore. A way was found to do this. If you recall, quasars are the most luminous objects of which we know and can be seen at very great distances. Thus before reaching us the light of a quasar travels through immense spaces. If it passes through gas clouds that are extended over vast expanses, even if they are extremely rarefied, some of the light will be absorbed, and the presence of the clouds will be revealed by absorption lines in the spectrum of the quasar. Many quasars exhibit absorption lines with red shifts markedly smaller than those of the emission lines originating from the quasars themselves. Since smaller red shifts correspond to regions of the universe that are expanding at lower velocities, and hence are closer to us, the absorption lines must be produced by matter interposed between us and the quasars. There is no doubt, therefore, about the existence of intergalactic gas. We can also determine which elements produced the absorption lines and the abundance of those elements. It turns out that the amount of intergalactic gas is negligible and, moreover, that it consists mostly of heavy elements. This shows that the gas is material processed by the galaxies rather than a residue of the original gas, which consisted essentially of hydrogen and helium. If any of the original gas exists, it is in minuscule quantities.

In sum, the galaxies were not born along with the universe; they were formed later and without waste of matter. But how? To answer this question astronomers have been pursuing two lines of research, both aimed at discovering what the galaxies were like around the time they formed and what were their first transformations.

The first approach starts from a simple proposition that has already proved very fruitful: When we observe galaxies with very large red shifts we see very distant galaxies. If the Big Bang occurred 15 billion years ago and the galaxies formed 10–20 million years later, when we observe galaxies at a distance of, say, 14.9 billion light-years, we see them as they were less than 100 million years after their formation. If we could see galaxies even farther away, we would be catching them at the time they were forming, or at least at the time they began to emit enough light to become visible. The idea is sound, and given telescopes powerful enough it can be done. The problem is, How do we recognize these galaxies? The best way would be to use the very large shifts in their spectral lines. But at that distance, even if they were much larger than our Galaxy or the Andromeda galaxy, they would look like faint luminous points, totally indistinguishable from the myriads of faint stars in the sky. Obviously it would be foolish to record the spectra of hundreds of normal faint stars, using up thousands of hours at the world's most powerful telescopes, in the hope of finding the spectrum of a newly formed galaxy. An alternative method has therefore been developed, namely, to predict what such an object should look like and then search for likely candidates.

This approach has been tried. E. E. Salpeter has calculated that young galaxies should look brighter because of the great amount of light contributed by the massive, short-lived stars that would have been very abundant early in the galaxies' lives. The results communicated by J. E. Tohline in 1979 also confirm that the first stars were very massive and hence very luminous. Through computer-generated models of three types of galaxies (spherical, elliptical, and disk-shaped), R. B. Larson has reached the conclusion that most galaxies formed from 100 million to a few billion years after the Big Bang. Furthermore, in young galaxies of these types the nucleus should be more luminous than the outer regions of the galaxy, in much greater proportion than in normal galaxies. At very large distances these young galaxies would look more like stars or quasars than normal galaxies, and this, as we said, would make their discovery next to impossible.

On the other hand, it is possible that we have already found forming galaxies without recognizing them as such. Actually we do know very luminous objects that have very large red shifts and therefore are very far away and very ancient. They are the quasars. At first, for these very reasons, they were thought to be the progenitors of galaxies. But some astrophysicists who tried to predict the characteristics of protogalaxies and the type of spectra they should exhibit have concluded that most likely it is not true. The reason is that the characteristics of the spectra predicted for newly formed galaxies are different from those of the observed spectra of quasars. Thus the quasars are neither protogalaxies nor the parents of galaxies. But they may represent a structural characteristic of newly formed galaxies, namely, a bright nucleus in a galaxy still very extended and too faint to be seen. This view is supported by Larson's theoretical model and seems confirmed by the existence of the BL Lacertae objects, the nuclei of Seyfert galaxies, and other peculiar galaxies. According to D. L. Meier and R. A. Sunyaev, the nuclei of these galaxies, which are less bright than quasars but have similar characteristics, could be the less conspicuous and closer cases (that is, quasars that with the passage of time have become a little less active and hence less luminous), while at greater distances there would be the brighter ones, because younger, in which the surrounding galaxies cannot be seen because of the distance. In fact, we have a few examples where observations with very sophisticated techniques have revealed a dimly glowing region, most likely a galaxy, surrounding a quasar. In conclusion, Meier, Sunyaev, and many others believe that quasars are not primordial galaxies, but the fact that large numbers of quasars are located just short of where we would expect to find the youngest galaxies (and therefore are slightly older) indicates that they may have developed in the central region of all or most galaxies right after their formation.

This line of research, based on the observation of ancient objects and on theoretical predictions concerning their formation and characteristics, has already borne fruit and may soon provide an answer to the question of when and how the galaxies were formed.

Let us turn to the second approach. It involves studying the galaxies as we see them now, measuring some quantities that should not change with time, and then working back from the current appearances and characteristics to the initial ones.

The quantities we assumed to be fixed were essentially two: mass and angular momentum. We have always thought, and still do, that the structure and appearance of a galaxy must change in time due to the condensation of gas into stars, the formation of clusters, and the variation in the distribution of stars, gas, and dust. However, since a galaxy is an isolated aggregate of matter, we assumed that its mass and total energy of rotation remained unchanged. With this assumption one could deduce the initial conditions from the current ones and learn how the various types of galaxies have formed and evolved.

Unfortunately, the findings of the past few years have raised a number of difficulties. In the first place, we have discovered that we do not even know very well the nearby galaxies whose current aspects we see —not even our own, the galaxy in which we live. The mass of a galaxy and its distribution with increasing distance from the center are determined from studies of the velocity of rotation. In numerous cases, as we reach the visible edge of the galaxy, which is as far as we can make observations, the mass does not seem to converge to a value that should be the total one. Hence the value we find for the mass is only a lower limit, and there is no way of knowing how much smaller it is than the actual one. Nor do we know where the extra mass is. Perhaps it is in the form of countless white or black dwarfs; perhaps it is in black holes, or simply in very faint stars, all of them distributed around the galaxy in an invisible halo due to its low luminosity.

This is just one of our difficulties. We have long known that there

Figure 83   NGC 6166, a large D galaxy at the center of a cluster, photo-graphed with the 1.26-meter Schmidt telescope on Mount Palomar. It is thought that this galaxy is progressively incorporating the others until the whole cluster will be swallowed up. [From *Palomar Sky Survey,* © 1960 National Geographic Society, reproduced by permission of the California Institute of Technology]

are interacting galaxies, in pairs or groups. In the 1950s they were often referred to as colliding galaxies. It was thought that they met by chance and, given the large distances separating their stars, they would go through each other with little or no damage. But recent computer simulations have shown that when two galaxies of the same size meet, they merge into one and the new one has different characteristics from its two components. Moreover, when minor galaxies arrive in the vicin-ity of a much larger one, they will be literally eaten up. In other words, we have discovered that there is a sort of "galactic cannibalism."

Cannibalism on a large scale may be taking place in clusters of galax-ies. In the center of a cluster there is often a radio-emitting giant gal-axy with an ellipsoidal nucleus and a vast external halo (figure 83).

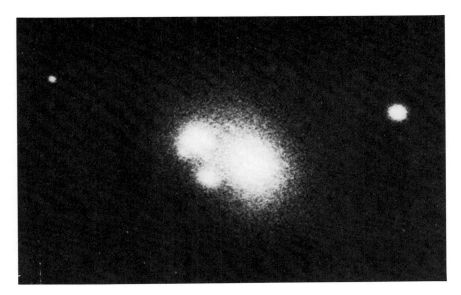

Figure 84   The central region of NGC 6166 in an enlarged view showing distinct nuclei that almost certainly are the largest remnants of galaxies that NGC 6166 has incorporated. [From *Astronomical Journal* 66 (1961):588]

Whenever there is such a galaxy, it is the largest in the cluster. Once it was thought that it had been that large from the beginning or that it might even be the remnant of the immense condensation that had given rise to the entire cluster. Today we think that the exact opposite may have happened, that the central galaxy was a galaxy a little larger than the others that began by incorporating smaller ones nearby (figure 84) and will go on to swallow larger and more distant ones—until one day perhaps the entire cluster will be reduced to a single mammoth galaxy of enormous mass.

And there is one more surprising fact. Until recently the so-called elliptical galaxies were believed to be either spherical or flattened into a lens shape by a high velocity of rotation about the polar axis. In 1972, however, F. Bertola measured for the first time the velocity of rotation of an elliptical galaxy (NGC 4697) to its faint outermost edge. He found a rotation velocity four times lower than those predicted from the models of J. R. Gott and R. B. Larson (figure 85). Starting in 1977, G. Illingworth obtained rotation curves for other elliptical galaxies,

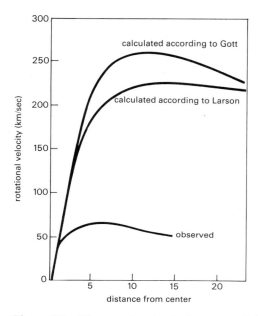

Figure 85   The rotational velocity curve of the galaxy NGC 4697 obtained observationally by F. Bertola and the curves predicted on the basis of the models of J. R. Gott and R. B. Larson. The distances from the center are expressed by assuming as the unit of measure the radius of the nucleus, that is, the distance at which the light intensity drops to half its central value. [From *Sky and Telescope* (May 1981)]

finding equally low velocity. This shows that a low velocity of rotation is not an exception but a fairly common characteristic of galaxies of this shape. Hence the flattening might not be due to the rotation of the protogalaxies from which they evolved.

In effect, theoretical and observational work has shown that an elliptical galaxy, when considered in three dimensions, may not be flattened in the shape of a lens, but rather shaped like an almond (oblate) or a cigar (prolate) (figure 86); in other words, it may be triaxial. Of course, it would be easy to tell whether or not an elliptical galaxy is triaxial if we could look at it from three different points in space, but obviously this cannot be done. Nevertheless, a way has been found to determine whether a galaxy is triaxial even from a single photograph taken from

Figure 86   NGC 5266, a galaxy surrounded by a dust ring only partially visible in the figure. According to F. Bertola, this galaxy, which could be considered a classical elliptical galaxy, might in fact be elongated rather than lens-shaped, constituting with others of the same type a class of cigar-shaped galaxies. [Courtesy F. Bertola]

the earth, our only observation point. In this case, too, we see it along a section, and figure 87 shows how its profile varies according to the way the triaxial galaxy is oriented with respect to us.

Thus we have discovered that there are even galaxies of this type. They cannot have taken this shape directly from the original proto-galaxies as an effect of rotation. Something must have happened later on, such as the fall of intergalactic material onto the galaxy or the capture of a smaller galaxy so as to alter or even destroy the original shape.

Finally, we cannot exclude that opposite phenomena may be taking place. Matter ejected from the nuclei of active galaxies may form minor galaxies that become satellites of the parent galaxies. Such processes

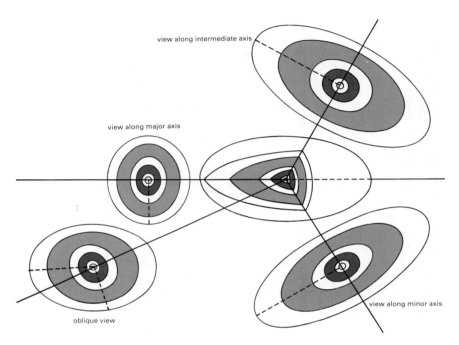

Figure 87   The four ways we can see a triaxial galaxy, shown right of center, partially cut in half. [From *Sky and Telescope* (May 1981)]

were proposed years ago by V. A. Ambartsumian and have yet to be proved groundless. On the contrary, H. Arp is discovering a growing number of strange galaxies or groups of galaxies that may be explained by invoking this or even more exotic mechanisms.

It seems clear from all this that in the course of their lives the galaxies can change markedly in appearance, shape, dimensions, mass, and rotation. Galaxies born at the same time at the beginning of the age of matter may no longer exist because they have merged, or have become part of larger galaxies, also born at the same time. The larger galaxies in turn may have changed as a result of incorporating smaller ones. Thus it is very difficult, if not impossible, to retrace the original galaxies from the current ones.

Yet this research, for the most part still in progress, has taught us something very important: With even a slight variation in one or two quantities, such as mass and dimensions, from the original clumps of gas, almost certainly very similar, that emerged at the dawn of the age of matter, there formed new shapes that subsequently continued to diversify. In other words, the galaxies, too, must have undergone the same process of transformation from the simple and uniform to the more complex and varied that we have observed everywhere and at all times. We do not know exactly when this process began. But we do know that once there was nothing in the universe but building blocks, all alike, while today space is filled with, so to speak, cities, towns, villages, and isolated houses, all structurally different, which live and change, renewing themselves like our earthly cities.

9 WHICH UNIVERSE?

Up to now, in all our journeys in time, we have gone back to more and more remote epochs; only for the sun and the earth have we ventured into the future in search of their destiny. But the solar system is an infinitesimal part of the universe; and the universe, in its entirety, moves toward a future of its own in which our future is just like an afternoon in the life of a man. The time has come to ask what is the future of the universe, as a whole and in its parts, and to try to discover if, just as it had a beginning, it will have an end, and if so, when it will end and how.

To answer these questions we must first understand in which universe we live. This does not mean that there are many different universes and that we must discover in which of them we happen to be. However, at least on paper, there is an enormous number of models of the universe constructed by the theoreticians. What we must try to find out is whether one of them corresponds to the only real existing universe by seeing, theoretically, how it is made and how it works over time. If all that concerns the past and the present corresponds to the real universe, we are entitled to think that its future will also occur as predicted by the model.

All we have learned about the past has revealed to us, fully or partially, how the universe has worked so far and of what it is composed. But we have never bothered to find out how it is made, for example, from the point of view of geometry. Yet is is important because, depending on its geometry, there can be different developments, that is, different futures.

THEORETICAL MODELS

The number of models of the universe that can be constructed theoretically is almost unlimited. Many have actually been elaborated, and additional ones are still being produced. We cannot examine them all, but it will be enough to discuss the solutions that best explain the fundamental facts discovered by observation, namely, the expansion of the universe, the way it expands, the existence and isotropy of the background radiation, the distribution of the galaxies, and others that we shall see.

FRIEDMANN'S SOLUTIONS
We shall start from a theory of gravitation (the main force governing the large-scale universe) that by now is well established: general relativ-

ity. The first to apply it to the entire universe, in 1916, was Einstein himself. Solving his equations, he found that the universe, whether finite or infinite, had to expand or contract. If the theory of relativity was correct, there were no other solutions—the universe could not be static. It was a shocking conclusion. At the time it was considered unacceptable, and Einstein himself rejected it. Instead, he postulated a cosmic force capable of neutralizing the effect of gravity and introduced into his equations a constant that would set things to right by making the universe static, as the good tradition wanted.

In 1922 A. A. Friedmann, a Russian scientist who had made contributions to the physics of the atmosphere but had never done work in cosmology, solved Einstein's equations anew. Friedmann assumed that matter in the universe was homogeneously distributed (practically anticipating the cosmological principle), and he ignored the "cosmological constant" introduced by Einstein. He found that the universe could be open or closed; in the first case, it would expand forever, while in the second case, after expanding to a maximum, it would start to contract.

His results were published in a brief scientific paper that did not receive much attention. In 1929, however, E. P. Hubble made a sensational discovery. He obtained the distances of several galaxies, and correlating them with their radial velocities, he found that the galaxies were moving away from us at speeds directly proportional to their distances. This meant[47] that the universe was expanding, just as Friedmann had asserted seven years earlier. It was a bitter moment for Einstein, who later admitted that the introduction of the cosmological constant was the biggest mistake he had made in his life. Friedmann did not live to enjoy the recognition he deserved because four years before the expansion of the universe was discovered he was dead of typhus at the age of 37. But in his brief paper he left us a momentous legacy. According to Friedmann, as we mentioned, there are three possible models of the universe, and one of them almost certainly corresponds to the real universe. Now it is up to us to discover which one.

THREE POSSIBLE UNIVERSES

Hubble's law describes how the recession velocity of the galaxies increases with distance and is expressed by a simple formula, $V = Hr$, where V is the velocity of recession, r is the distance, and H is a constant known as the Hubble constant. This constant tells us how the universe is expanding by means of a number that expresses the expansion velocity in kilometers per second for each megaparsec of distance. (1 megaparsec = 1 million parsecs = 3,262 million light-years.) It is calculated from the recession velocities and distances observed for a number of galaxies. In theory one galaxy should suffice, but in practice we must use many of them because the number we are seeking is the velocity with which the universe—all of space—is dilating. The galaxies thus act as indicators of the velocity of space, much as the dots painted on a swelling balloon indicate by their displacement how the volume of the balloon is increasing. In the case of the balloon it is all very simple because the dots are painted on it and therefore none of them moves from where it was painted. But the galaxies do move, both in groups and individually within each group. Hence if we consider only one galaxy, we run the risk of measuring a velocity that is a combination of the expansion velocity of the universe and the velocity of the galaxy's own motion in space. Thus to obtain an accurate value of the Hubble constant we must calculate it from the largest possible number of galaxies situated at various distances from us. And even that is not good enough because the effect due to the motions of the galaxies, including ours, within the universe is not known very well. As a result different authors have obtained very different results, and particularly of late the exact value of the Hubble constant has become very controversial.

The problem is serious because from the Hubble constant we derive the age of the universe, among other things. In itself, the procedure is simple. Let us assume that the universe has always expanded and will continue to expand at the same velocity as today. We denote by R the current distance between any two galaxies. It is obvious that due to the expansion of the universe, the distance between the same two galaxies will continue to increase in the future, while in the past it must grow smaller and smaller until it drops to zero at the beginning of the universe. If we plot the increase of R (distance between the two galaxies) with the passage of time, assuming the universe has always expanded at the same velocity, we obtain a straight line (figure 88), which will be more or less inclined to the time axis, depending on the velocity of expansion. The value of R is called by cosmologists the *scale-factor of the universe*, because it tells us at what rate the universe is expanding. Figure 88, in which R is plotted against cosmic time, shows that R becomes equal to zero when the straight line meets the time axis. That is the

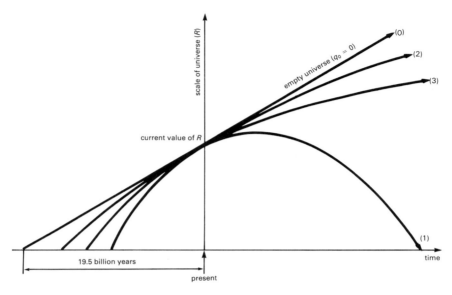

Figure 88   How the universe expands with the passage of time. The diagram illustrates the three solutions found by Friedmann: (1) closed, (2) open, and (3) flat universes. As the figure shows, in the first case, the distance R between two given points—for example, two galaxies—after reaching a maximum will start to diminish again, and the universe will end in a point as it had begun; in the second and third cases, instead, the distance will continue to increase and the expansion is destined to go on forever. [From *Mercury Magazine*, © 1978 Astronomical Society of the Pacific]

moment when the entire universe was concentrated at a point. The interval of time measured between that instant and today indicates the age of the universe. The value of this interval is simply the inverse of H.

   All this would be all right if, once it had received the initial push at the moment of the explosion, the universe kept on expanding at a constant velocity, which would be the case if there were no forces tending to slow down the expansion. But there certainly is one force that counteracts the expansion, and that is the force of gravity. Hence the universe must have expanded more rapidly in the past; this means that it is younger than the age deduced from the line of constant expansion. But how much younger? Actually there are three cases, corresponding to different types of universe, that represent the three solutions found by Friedmann.

We know that there was an initial force that provided the enormous push for the expansion that is still going on. We do not know its source or its magnitude, but with regard to the latter it is easy to conclude that there are only three possibilities: (1) the energy of the universe is not sufficient to overcome the decelerating effect of gravity; (2) the initial push was exactly that sufficient to make the universe expand to infinity in an infinite time; or (3) the initial push was greater than the latter value and there is even more reason for the universe to keep on expanding to infinity in an unlimited time.

In all three cases the expansion tends to slow down; that is, it is subject to a deceleration expressed by a parameter that the cosmologists denote q_0. But it can slow down at different rates, more or less rapidly, and consequently the deceleration can have three different outcomes: (1) If the value of q_0 is greater than $\frac{1}{2}$, the universe will eventually come to a halt, reverse itself, and contract back to a point; (2) if q_0 is exactly equal to $\frac{1}{2}$, in an infinite time the expansion will slow to a stop, but will not reverse; (3) if q_0 is less than $\frac{1}{2}$, the deceleration is not sufficient to stop the expansion, which will continue forever. Thus the different values of q_0 express the three models of the universe proposed by Friedmann (figure 88). They represent three possible futures of the universe, and something else, too. Which of the three is the real one we could know right now, because the three models we have followed in time correspond to an intrinsic property of the universe—its geometry.

As demonstrated by general relativity, the presence of a mass causes space to curve. The curvature becomes more pronounced the greater the mass of the body that causes the curvature is in relation to its volume; in other words, the higher the density (mass/volume), the greater the curvature. The total mass of the universe is proportional to its density, which diminishes continually because of the expansion. Let us take a given instant in time. If the average density of matter is greater than a certain critical value, space curves to the point of closing onto itself. The paths of all bodies, from particles to galaxies, and even the paths of light rays, will curve accordingly, and everything will be trapped in a universe that for this reason we call *closed*. If the density is lower than the critical value, space curves the opposite way, so to speak, and is called *open*. At the boundary between these two groups there is a value of the density whereby space-time has a configuration at the limit between the two; this is the *flat* universe.

These concepts of the geometry of the universe are not at all easy to understand. We are three-dimensional beings, and also the space in which we live, at least in our neighborhood, appears to be three-dimensional. For this very reason it is difficult for us to understand what happens in this space, let alone how its large-scale properties change. On the other hand, we can easily understand a geometry with one or two dimensions less. In other words, while we have troubles seeing a three-dimensional body, like ourselves, we have no difficulty in seeing and comprehending in all its properties what is drawn on a plane or a straight line. Thus we shall try to envisage the previous concepts in relation to a two-dimensional space that can curve in the third dimension.

Now everything becomes clear (figure 89). Let us start from the third case, the limit, or the flat universe. In two dimensions, this type of universe can be represented by a plane, which, as everybody knows, is infinite, and where Euclidean geometry, the geometry one learns in junior high school and the only one known to most of the people in the world, holds everywhere. On that plane two flat beings (devoid of thickness) will find that the sum of the three angles of a triangle equals two right angles. This result will hold always and everywhere, whatever the shape and size of the triangle. If we now move the flat beings onto a sphere, they will find that this result and all of their geometry hold only as long as they keep to a fairly limited region; but if they make large-scale measurements, they no longer will obtain the same results. This is reasonable; they are no longer living on their infinite plane but on a finite surface, closed onto itself in a third dimension of which they are completely unaware. Finally, there is the case of a surface that, contrary to the spherical surface we may regard as positively curved, is curved negatively, that is, outward. Unlike the surface with a positive curvature, this surface is infinite.

The drawings in figure 89 illustrate the three types of universe, but of course you must remember that these are only examples lacking a third dimension. We can vaguely imagine the two universes, open and closed, predicted by Friedmann through the first—the flat Euclidean—one. Figure 89 shows it as a plane, but in reality it has three dimen-

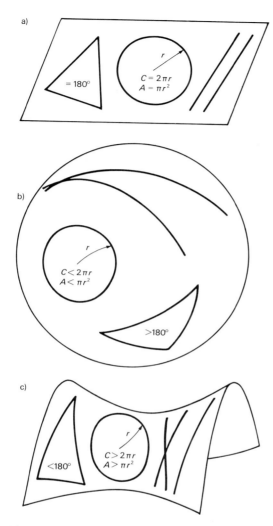

Figure 89 Two-dimensional analogs of Friedmann's three types of universe. (a) the flat universe in which Euclidean geometry is always valid—two parallel lines will keep the same distance between them throughout the universe; (b) the closed universe, which in two dimensions corresponds to the surface of a sphere—here two parallel lines converge; (c) the open universe, in which two parallel lines diverge. In all three cases, which are mere fictions to show what can happen in our universe by observing what happens in a space with a dimension less, the galaxies (like any body belonging to each of the three universes) must be imagined flat, like very thin films devoid of thickness. Naturally in (a) and (c) the edges have been drawn only because the page of the book is not infinite, as the corresponding universes in fact are. [From *Le Scienze* (July 1976)]

sions. Something of the kind must be imagined for the other two universes whose properties the scientists can calculate but which we cannot visualize, even though one of these two may be the universe in which we live. But we can live happily in it just the same because, thanks to our smallness, we are very much like the two flat beings unaware of having been moved onto a sphere, that is, in a world curved in a dimension they do not know.

SUMMARY AND CLARIFICATION
Going quickly over what we have seen, we can make a synthetic picture of the possibilities that today are held to be most valid. Table 7 shows the three main models of the universe. They are (a) closed universe, finite, closed onto itself—it has a deceleration parameter q_0 greater than ½, and it will expand up to a certain time, when it will stop and then will contract back to a point; (b) open universe, infinite and in eternal expansion having a deceleration parameter between 0 and ½; (c) flat universe, a universe for which Euclidean (everyday) geometry holds everywhere, whereas for the other two it does not hold on a large scale—for this universe q_0 is exactly ½ and in an infinite time the expansion will slow to a halt but will not reverse.

At this point we should stop a moment and consider some aspects of these models that are not evident but are very important, starting with the age of the universe in all three cases. Let us go back to figure 88. By adopting the most recent value of the Hubble constant calculated by Sandage and Tammann, we find that if the expansion had not slowed down, the universe would be 19.5 billion years old. But we know that it must have slowed down; hence the actual age of the universe is certainly less. If the universe were flat, its age would be 13 billion years; if the universe were open, its age would be between 13 billion years and the value we have excluded, 19.5 billion years; and for the closed universe it would be less than 13 billion years.

Another point that needs to be clarified concerns the radius of the universe. When we speak of a Big Bang, instinctively we are apt to think that the universe began from a point, as suggested by figure 82. In effect, this is literally true only for the closed universe, which at the beginning of time was schematically reduced to a point where the scale-factor R was zero and the density infinite. It is a situation inconceivable to us, and on the other hand, nature, through the laws of physics, has

Table 7   Models of the universe[a]

Characteristics		Open	Critical	Closed
Density parameter Ω	$\dfrac{\text{actual density}}{\text{critical density}}$	$\Omega < 1$	$\Omega = 1$	$\Omega > 1$
Deceleration parameter q_0	$\dfrac{\text{distance}}{(\text{velocity})^2}$	$q_0 < \frac{1}{2}$	$q_0 = \frac{1}{2}$	$q_0 > \frac{1}{2}$
Space geometry		Hyperbolic (negative curvature)	Flat (zero curvature)	Spherical (positive curvature)
Future of the universe		Perpetual expansion	Perpetual expansion	Final collapse
		╱	╱	⌒

a. From *Le Scienze* (July 1976).

denied us access to it by the barrier of 10^{-43} second, which, if it prevents us from knowing the most important moment in the life of the universe, also prevents us from going mad trying to understand something that, at least for now, is beyond our comprehension. Such a situation is known as a *singularity*.

In the open model, instead, the universe is infinite and has always been so. In this case, the Big Bang takes on a different connotation, namely, an explosion that occurred at each point of an infinite universe. Figure 82 does not reflect the concept. On the other hand, if the universe is closed, the idea of a continuous expansion of space from a point suggested by the figure is correct, and the illustration can be taken literally.

THE FUTURE OF THE UNIVERSE

Now we are ready to embark on our last journey in time: the journey into the future of the universe. Actually it will not be one journey but three because there are three possible futures according to our current knowledge, and not knowing yet in which universe we live, we cannot tell which one will come to pass. The three possible futures do not correspond to Friedmann's three types of universe. For two of them, open and flat, the future would be the same, while for the closed universe there are two alternatives, one of which would involve the past as well.

SLOWDOWN AND COLLAPSE

If the universe is closed, its expansion will progressively slow down and eventually will come to a halt. For an instant the universe will stand still. Then it will start moving again, but in the opposite direction, contracting faster and faster, then collapsing toward the point from which, long before, all that had been built over billions of years had originated. We do not know when the universe will stop expanding. The time and whether the universe stops expanding at all depend on a particular value of its density at a certain instant, known as the *critical density*. Let us say that its density is twice the critical density. In this case, the universe will expand to twice its present size and then will start to contract. It might be exciting to be there at that moment, but nobody on earth will live to see it because by that time—tens of billions of years from now—the solar system will consist of frozen planets orbiting an almost dead star. But we do not have to be home to watch the show. We can be anywhere in the universe—in a space colony, if you like.

No matter where we are we would see the same events, which at first would not be at all spectacular. Today we see all the galaxies recede at velocities that increase with distance. In the instant of cosmic time in which the expansion stops we would still see the more distant galaxies recede because the light we receive at that moment is the light that left those galaxies millions or billions of years earlier, when the universe was still expanding. We would only see them receding at lower speeds because the universe was already slowing down. A few million years after the halt, however, we would see some galaxies stop—the ones closest to us, which tell the story after the relatively short time it takes their light to reach us. But meanwhile, of course, the collapse has started. Billions of years go by and nothing much seems to be happening. The universe, which until the moment it stopped had continually grown, returns to its current size and keeps on shrinking, but is still so vast that each galaxy can continue to lead its normal life. Only an astronomer who observes at the telescope and measures the radial velocities of the galaxies will be aware that something strange is happening, and even if he knows nothing about the past, he will understand that what he sees does not bode well for the future. Measurements of the radial velocities show that the nearest galaxies are approaching us; others, farther away, are standing still; and those beyond are moving away from us. Again, it is the nearby galaxies that tell the real story. For the others there is the usual time lag due to their distance, which shows them as they were billions of years earlier, when the expansion had just halted, and still earlier, when it had not yet halted.

More time passes; then quickly the situation becomes desperate. About 100 billion years have gone by since the universe began, and if you think how much has happened on the earth in just 4 billion years, you can imagine how much will have happened in the whole universe in a time 25 times longer. It is all about to end. In the space of a few million years, which might seem like a very short time if our biological processes were slower, the drama is consummated. Telescopic observations show that the galaxies have moved much closer and that all are converging toward us at tremendous speeds. This does not mean that they all have something against us. On the contrary, if somebody observed the sky from any of the galaxies that seem to threaten us, he would see our own plunging toward him along with all the others. We would feel that we were at the center of a universal collision, just as today we feel that we are at the center of a universal expansion. Regardless of appearances, the terrible truth is that the world is getting tighter and tighter.

Something else has happened. Due to the contraction of the universe, the background radiation, which had dropped to 1.5°K at the moment the expansion stopped, has risen to 300°K. Everywhere in the universe the sky has become as hot, day and night, as it is now during the day. The heat does not come from any particular body, like the sun, but from the entire sky, because 300°K is the temperature of the universe. And the heat continues to rise until it removes the danger of galactic collisions by more destructive means, by destroying and pulverizing everything before the galaxies have time to collide. Temperature and density increase faster and faster in the last thousand years, the last years, the last seconds, creating a cauldron where the whole universe melts along with its present and its past. Then even "melted" matter is gone, and there is only energy. One more instant and everything disappears into a point. The universe is finished.

AND THEN?

It is all over—like the life of a human being, like the life of everything that is born. Having been born, the universe too had to die. But are we quite sure that it is finished? After all, we did not follow the universe to the very end, just as we had not been able to see it at the instant it was born. There are always those 10^{-43} seconds—the very first and the very last—that not even theory can penetrate. Perhaps the universe does not end in a point of infinite density where space and time cease to exist. Perhaps it did not begin with the creation of everything from nothing.

In the last thousandths or billionths of a second of the collapse, in a very small space, the temperature, density, and energy are enormous. In the very last instants, perhaps in the infinitesimal fraction of a second we cannot explore, there might be a sort of rebound and the huge amount of energy generated by the collapse might serve to produce another explosion. In other words, there might be a new Big Bang, and everything would start all over again for another 100 billion years. But it is not clear that everything would happen in the same way. In the past few years two schools of thought have emerged that hold different views about the first moments of the universe, and at least one of them rules out an exact repetition of the previous universe.

According to that view, the universe initially went through a *chaotic,* discontinuous state. This chaos would have erased any trace of a possible previous universe and any conditions and physical laws prevailing in it. Thus everything would have started all over again following entirely different routes, except for chance repetitions. Cosmological theories based on a chaotic early universe eliminate the problem of having to discover its initial conditions because it cannot be done—it is impossible to understand the structure and laws of a universe ruled by chaos.

This theory, however, meets with a serious difficulty. We know that today's universe is homogeneous and isotropic, and we have evidence that it has been so since epochs very close to its origin, possibly since the first 10^{-35} second of its existence. In no way can these properties be reconciled with chaos. On the contrary, they characterize a universe that should retain such properties even, and particularly, at the beginning because of its smaller size and the greater simplicity of its structure. This type of universe is called *quiescent.*

If the universe was initially chaotic, when, how, and why did it become the quiescent universe that we know today and that may have been so as early as 10^{-35} second after the Big Bang? There are no easy answers. Furthermore, B. Collins and S. Hawking have demonstrated that a universe not perfectly regular is unstable. In other words, a universe initially irregular would become more and more irregular.

In any event, whether beginning and end be chaotic or quiescent, a closed universe, and hence subject to an expansion and a collapse, could start over many times. In the chaotic case it could start over an infinite number of times, but each time it would be a new, autonomous universe, whereas in the case of the quiescent universe there could be ties between the various cycles.

A cyclical universe that begins, ends, and begins over and over again is known as an *oscillating universe*. Thus the closed universe can be of two types: a unique universe which has a beginning and an end, and an oscillating universe, which starts and finishes we know not how many times. It may have already exploded and collapsed an infinite number of times, and it may continue to do so forever.

ETERNAL EXPANSION

But the universe may not be closed. Suppose it is open. Here our scenarios for the future border a little on science fiction, partly because of the important role that could be played by black holes (whose existence, however, has not yet been confirmed), and partly because of the developments that one could imagine in living beings, who, by understanding and acting in the universe, can make a contribution not only in life, but also in thought and above all in action so as to counteract, direct, or at least influence the processes at play in the universe.

In an open universe the galaxies will continue to move farther away from one another but will retain their individuality. The expansion of the universe does not increase the volume of the galaxies, nor in itself does it alter their structure—in the same way that inflation does not increase the value of money. Thus, aside from the increasing global dilation of space, the future of this universe must be traced through the evolution of the galaxies.

It has been calculated that between 100 billion and 100,000 billion years from now all the stars in all the galaxies will be dead; that is, they will be reduced to white dwarfs, neutron stars, and black holes. In addition to these remnants of the stellar world there will be the remnants of the matter that had been cold for a long time, such as planets, meteorites, and cosmic dust. Along with dead stars, all these bodies that today we consider dead, compared to stars and active planets, but actually have a life of their own, will continue to constitute galaxies that for a while will remain structurally intact, though nearly spent and dark. In time, however, much of the material in each galaxy will be redistributed in a new way. There no longer will be associations, star clusters, or the large aggregates of the spiral arms. A large, dense nucleus will form, and eventually its massive center will become a black hole of about 1 billion solar masses, while most of the other bodies will be expelled outward into intergalactic space. According to J. N. Islam, all this will occur in about 10^{20} years—an enormous number, considering that it is equivalent to 100 billion billions. We are beginning to deal with very long times, far exceeding the 100 billion years predicted for the entire life of a closed universe.

If during the expansion of the universe the clusters of galaxies were to remain gravitationally bound, their galaxies, along with the bodies that escaped from the formation of galactic black holes, could form supergalactic black holes of 100 billion to 1,000 billion solar masses.

Contrary to what was formerly thought, black holes are not eternal. In 1975, by taking into account quantum mechanics, S. E. Hawking found that a black hole must emit electromagnetic waves and neutrinos; in other words, it would not be perfectly black, and because of this emission, little by little it should evaporate. This process is extremely slow—a galactic black hole should last 10^{90} years; a supergalactic one, 10^{100} years. Thus in an inconceivable, but finite, number of years all the black holes in the universe will have disappeared, and in place of the dissolved galaxies there will be neutron stars, black dwarfs, planets, and rocks, all afloat in a space that continues to expand, growing emptier and emptier.

But not even solid bodies will be there forever. Due to a physical process that we shall not discuss (tunnel effect), even the most rigid material will undergo a kind of dissolution in about 10^{65} years. Neutron stars and black dwarfs will share this fate, turning into black holes in $(10^{10})^{76}$ years—a time so long that it seems to merge into eternity. But it is not eternity. Eventually even this long period of time will come to an end. When the last generation of black holes has disappeared, nothing will be left of the universe we know today, so rich and varied, but electromagnetic waves and neutrinos, increasingly more diluted in a space that will continue to expand for all eternity.

But this picture is not complete, and perhaps will never materialize, because we have not yet considered something very important that is also part of the universe—living beings.

As matter is transformed to the point of turning into radiation and light particles in an ever expanding universe, the living world also will be transformed. Of course, it is unthinkable that in those distant epochs there should still be life on earth, long since gone, or that there should exist organisms similar to those that populate our planet today. But living beings might still inhabit a countless number of worlds. Considering how rapidly life and intelligence have progressed here on earth, should biological evolution continue at the same pace, one could foresee that in the distant future the universe will be populated by beings so much more advanced than we are that one cannot even use the familiar comparison between amoeba and man. It is also possible that terrestrial life itself, not necessarily human but certainly very advanced, will migrate from our planet before its destruction and, once transplanted on other planets or in space colonies, will continue to contribute to the evolution of universal life. In any event, even if life as we know it should be totally extinguished, what happened on the earth will not have been a meaningless experiment but rather a necessary step in the transformation of cosmic matter. One cannot climb up a ladder without building the first rungs. And the advanced beings surviving elsewhere in the universe also will have evolved from primitive life forms as occurred on the earth.

Naturally this can happen only if there is neither an inversion in evolution nor a general extinction of life all over the universe. For life to go on, some form of energy is indispensable. That is the basic necessity that the nonliving world must furnish to the living one. Today energy is provided almost exclusively by the stars. In 100 billion to 100,000 billion years from now almost certainly all the stars will be extinct, but long before that happens technically advanced civilizations may be obtaining energy from black holes.[48] Some civilizations might fail to find black holes in the darkness of space, but as long as there are stars burning somewhere, these civilizations may be able to survive by moving close to them—say, by means of nuclear propulsion. Meanwhile, a galactic black hole will form, which will be able to furnish plenty of energy for a very long time to all the civilizations that have moved into its vicinity. A galactic black hole of 10 billion solar masses would have a radius of only a few light-days. All the galactic civilizations that wanted to exploit its energy, coming from the most disparate and distant places, would find themselves circling about it in a space small enough to permit communications and exchanges. Thus the black hole would become like the oasis that serves as a meeting place for peoples of different races and cultures from the various parts of the desert. If this happens, galactic civilizations will be able to last another 10^{90} years by exploiting the energy of a galactic black hole and as much as 10^{100} years by using a supergalactic one.

All this may be pure fantasy, and it certainly is if black holes do not exist. In any case, even by exploiting black holes, life could survive for an inconceivably long but still finite time. Thus we come to the fundamental question: Can living beings survive forever with a fixed and limited amount of energy? It seems impossible, but perhaps it is not. F. J. Dyson has an intriguing theory on the subject. He starts from two necessary assumptions, namely, that we already know all the fundamental laws of physics and that they will not change in the future. This said, he notes that it has taken times of the order of 10 billion years for primordial matter to evolve into *Homo sapiens*—specifically, times of the order of 1 billion years for the development of a phylum, 100 million years for the development of a class, 10 million years for the development of a genus, and 1 million years for the development of a species (for example, ours). This is what we have seen in our journey through the earth's past: Evolution has been proceeding at an accelerating pace toward life forms of increasing complexity, intelligence, and sensitivity. Most likely, it is also a sample of what is occurring on a number of worlds that would be very large by any standards even if it were a minute fraction of all the existing celestial bodies.

If so much has happened in the 15 billion years from the Big Bang to the present, it is almost impossible or awe-inspiring to think how life will evolve in the next 15 billion years. One can entertain all kinds of fantastic possibilities, but Dyson asks a more basic question: Suppose we could build a replica of a brain with the same structure but different materials; would the replica work as well as the original? And more to the point: Does the basis of knowledge reside in matter or in structure?

If the answer is "matter," then life can continue to exist only as long as there is a warm environment with liquid water and a constant supply of free energy to sustain a stable metabolism. In this case, since each galaxy has only a limited supply of energy, the duration of life also would be limited in time. On the other hand, if the answer to Dyson's question is "structure," then it is conceivable that in the future life may evolve from a flesh-and-blood structure to something in the nature of a computer or a sentient, self-aware interstellar cloud like the "black cloud" of Hoyle's famous novel. Naturally, such beings also would have to replicate themselves by handing down their hereditary traits.

Pursuing this line of reasoning, Dyson suggests that a scale law such as exists in physics could apply in biology. In this case, he has demonstrated that with recourse to intermittent hibernation such a life form could survive for an infinite subjective time on a finite amount of energy. According to Dyson's calculations, a society as complex as the present human race could live forever on the amount of energy that the sun currently emits in just eight hours. Such a society need not have a limited memory. By using the type of memory of an analog computer it could have a memory of ever increasing capacity. Thus, if the universe is open and continues to expand, the living world could continually renew and cause to last forever the material world from which it was born that otherwise, even though in a very long time, would have to end.

In the closing page of his book *The First Three Minutes,* S. Weinberg writes, "The more the universe seems comprehensible, the more it also seems pointless." This may be true when we consider a universe of galaxies, stars, and desert planets. But that is not all there is to the universe, and in predicting its future one cannot ignore the fact that it consists of life and intelligence as well. This is the starting point of Dyson's reasoning. His theory may not correspond to reality, but the fundamental consideration remains valid, and even if the global picture should be different, it would still correspond to a universe ruled not only by the laws of physics but also by those of a living world in continuous transformation and evolution. Thus any prediction that ignores biological evolution can only be incomplete and distorted.

THE TWO PROSPECTS

When we discussed the fate of the closed universe we saw that it should last 100 billion years. This result follows from the fact that we had attributed to it a density twice the critical one. If the value of the density is higher, the universe will last less. In any event, its duration is independent of the level of development reached by stars and planets. During the final phase of the collapse, a closed universe would drag along dead stars and planets (as the sun and earth will be by that time) as well as active stars and planets of various ages, worlds where life had just started to flourish or was about to achieve a high degree of development. What is so extraordinary about it is that after the brief initial eras, the universe has undergone a continuous evolution that is independent of the fact that it expands. Hence the collapse would be likely to destroy the greatest system of which we know at the moment of its highest level of development. Of course, none of this would apply if the life of the universe resembled that of a living being, which after the ebullience of youth and the stability of middle age begins to decay in body and often in mind as well. In this case, having lived 15 of the 100 billion years predicted as its life span, our world would correspond to a youth at the height of its strength and intelligence.

It would be a different story altogether if the universe expanded to infinity and for all eternity. In an open universe nothing would cut short the process of development or decay of the individual components. The ultimate result would not depend on the behavior of the universe as a whole; on the contrary, it would always be the sum total of the individual contributions of each star, planet, and civilization. In a closed universe the continuous refinement of matter in forms ever more complex and incorporeal would have a very long time at its disposal, but one day, inevitably, it would be curtailed. In an open universe times would be enormously longer, and if Dyson is right, the quality and destiny of the universe would be determined by its constituent matter and its potential—not only the unconscious potential of inorganic matter but also the conscious potential of a matter that lives, thinks, and acts accordingly.

In other words, in an open universe the contribution of each individual is meaningful—indeed, necessary—and no contribution is lost, from the mating of two reptiles in the Jurassic to the bean soup made by an old woman in a remote hamlet. For each contribution is a step on an ascending ladder; broken or shaky steps will fall off, but nonetheless, they are necessary because the higher steps could not exist without them. In a closed universe the individual components resemble students who are given some time to write a composition, but whose papers are taken away when the time is up, even those just started, and all are thrown into the fire unread. Does this mean that our choice is between a democratic universe and a cold-blooded dictatorship? Not exactly. The universe is neither kind nor cruel. It is what it is by virtue of its geometry, and nothing can be done to change that. But by discovering which type of universe we live in we can give a different meaning to our lives.

If the universe is closed, the awareness that everything will end, no matter how far into the future, would remove any sense of responsibility toward the cosmos and make one's life closed and an end in itself. The meaning of such a life and the way to live it are poignantly expressed in Ecclesiastes, written more than two thousand years ago: "Go thy way, eat thy bread with joy, and drink thy wine with a merry heart; for God now accepteth thy works. Let thy raiments be always white; and let thy head lack no ointment. Live joyfully with the wife whom thou lovest all the days of the life of thy vanity, which he hath given thee under the sun, all the days of thy vanity. . . . Whatsoever thy hand findeth to do, do it with thy might; for there is no work, nor device, nor knowledge, nor wisdom, in the grave, whither thou goest." This grave—Sheol, a place without light or joy where the dead abide as shadows (shadows in the darkness, that is as nothing in nothingness)— is the fate that awaits the closed universe, the ultimate singularity that swallows everything into a point without dimensions.

But if the universe is open, if matter will always tend to a refinement and a spiritualization for which there are no limits, then our life not only is suffered and enjoyed while we are living it but assumes a cosmic significance unlimited in time because everything we have done, even if infinitesimal compared to the space and time of the universe, will have a meaning for all eternity.

THE REAL UNIVERSE

We have discovered that knowing the type of universe in which we live is also important in order to understand the role played by the living world and ourselves. Hence it is necessary and pressing to discover which model of the universe corresponds to reality. Various methods have been devised to arrive at this crucial discovery, but at the moment, unfortunately, there are practical difficulties with all of them that prevent us from reaching a definite conclusion. It is hoped that this is only a temporary setback. Our means of observation continue to improve, and scientists are building instruments that in a few years could give us the great answer. Furthermore, we have obtained some results that, though inconclusive, are worth mentioning.

Let us go back to figure 88. From the current rate of expansion of the universe expressed by the Hubble constant, we find that in the absence of deceleration, that is, if the universe had always expanded at the same speed, it would have originated at a certain epoch. Taking the value of the Hubble constant calculated by Sandage and Tammann, that epoch falls 19.5 billion years ago. But as we have seen, there must be a deceleration. Initially, therefore, the expansion must have been more rapid, in different measure depending on the type of universe. Thus the three models give us different dates for the birth of the universe, all calculated on the basis of the Hubble constant. If we could discover the date of birth of the universe with a method that is independent of the Hubble constant, by comparing our observational result with the dates calculated for the three models we could determine which type of universe we live in.

One way is to find the age of the oldest stars—for example, those in the metal-poor globular clusters. The stars cannot be older than the universe. Thus, if their age is greater than that corresponding to the intersections of the curves 1, 3, and 2 with the time axis in figure 88, any model of the universe that falls below can be ruled out. In particular, adopting 19.5 billion years as the age of a nondecelerating universe, if the age of the stars is between 13 and 19.5 billion years, the universe is certainly open, while if it is below 13 billion years, the universe may be closed. This second case does not exclude an open universe because one day we could find stars or other objects of greater age, but until such time it would favor the model of the closed universe. The trouble with this method is that it requires an accurate determination of the Hubble constant because it is this constant that determines the age limit and subsequently the dates of birth of the universe according to the various models. At present, unfortunately, the value of the Hubble constant is very much in question. However, if we assume that the age of the oldest globular clusters is 17 ± 2 billion years (A. Sandage, 1982), and that the value adopted for the Hubble constant is the one that today gives the highest value for the age of the universe, the results obtained with this method tend to rule out that the universe is closed.

Another method is based on the determination of the mean density of the universe. We take a number of clusters of galaxies and calculate their masses from the motions of the component galaxies. In this manner we find the visible and invisible mass contained, on the average, in a large number of clusters. From measurements of their apparent diameters and distances we obtain the clusters' volumes. Dividing mass by volume, we obtain the density of each cluster. On a very large scale, clusters and superclusters can be considered to be fairly uniformly distributed in the universe. To find the mean density of the universe we must also take into account the mass of the matter that may be scattered between clusters, but until recently it did not seem to be a significant amount. From estimates of the density made with this and other methods we have found values 10–100 times lower than the critical density, again indicating that the universe is open. There is a problem, however. We had always thought that the mass of the neutrino was nil, or if not nil, negligible. Recently we have begun to suspect that it may not be. If you recall, in the first instants of the Big Bang there formed an immense number of neutrinos that are still at large in the universe —in galaxies, in clusters of galaxies, and in the space between clusters. Even if the mass of each neutrino is small, multiplied by a very large number it adds up to a mass that might push the mean density over the critical value, thereby closing the universe.

The most direct way to discover the type of universe we live in is to use Hubble's law, which says that the greater a galaxy's distance from us, the higher is the velocity with which it seems to move away from us. This is only the way it appears to us, as shown by the example of the balloon expanding at a constant velocity.[49] If the expansion is slowing down, when we observe the galaxies, i.e., the universe, in epochs closer to the Big Bang, we should find that the velocity of expansion was higher then than it is now. From the difference in the velocities, we can obtain the value of the deceleration parameter q_0 on which the type of universe is dependent. To observe the galaxies in the most remote epochs we need only look for the most distant galaxies, whose light, emitted eons ago, reaches us now, telling us what the universe was like then and how it expanded.

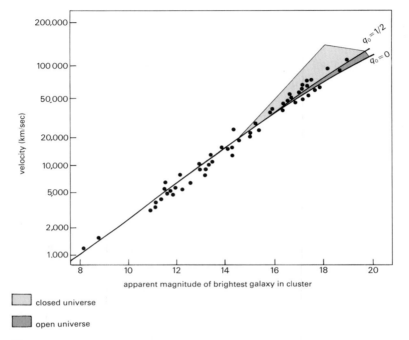

Figure 90   A graphic representation of Hubble's law for Friedmann's three models of the universe. The curves coincide for the nearby galaxies, which appear brighter. Beyond a certain distance the curves bend upward within the lightly shaded region if the universe is closed, fall in the darkly shaded region if the universe is open, and fall on the line $q_0 = 1/2$ for the flat Euclidean universe. By plotting observed galaxies in this theoretical diagram, depending on where the corresponding points fall we can determine which type of universe we live in. [From *Mercury*, © 1978 Astronomical Society of the Pacific]

In figure 90 the radial velocities measured for the brightest galaxies in clusters of increasing distance are plotted against their apparent magnitudes, which become higher with distance because the galaxies grow fainter. Also shown are the two curves corresponding to the decelerations $q_0 = 0$ and $q_0 = 1/2$. Observe that up to a certain distance the theoretical curves coincide, diverging only around magnitude 18. This says something we already knew, namely, that as long as we consider it on a small scale, the universe appears the same whether it is open, closed, or flat. But at a certain point the curves diverge. Thus if the points representing the most distant galaxies fall on a curve comprised between those corresponding to $q_0 = 0$ and $q_0 = \frac{1}{2}$, it means that the universe is open; if they fall on the curve of $q_0 = \frac{1}{2}$, the universe is flat, or Euclidean; if they fall on an up-turning curve corresponding to $q_0 > \frac{1}{2}$, the universe is closed. The distribution of the points in figure 90—and in particular, the positions of the last three points at the top—seem to indicate that we live in a flat universe, but perhaps this conclusion is too hasty. At the observed magnitude (a little less than 20th), the curves are still too close and the representative points too few to give conclusive results.

There is an additional difficulty. For this method to be applicable, all the galaxies used in the graph must have the same absolute magnitude, that is, the same intrinsic brightness. But we cannot be sure that younger galaxies (the most distant ones) had the same intrinsic brightness as the older galaxies we see nearby. Perhaps newly born galaxies were much richer in young stars and supernovae and therefore, seen at the same distance, would be more luminous. On the contrary, galaxies may be more luminous today than they were at an early age, having cannibalized some of the smaller components of the cluster. It is also possible that both things are true, and the opposing effects may have neutralized each other. If we can solve this problem theoretically or observationally, that is, if we can find how the intrinsic brightness of galaxies varies with age, we also may be able to solve the fundamental problem in a short time. In a few years the Hubble Space Telescope will start operating. This new observatory will make it possible to obtain the spectra of galaxies as faint as the 25th magnitude, 100 times fainter than any observed to date. Looking at figure 90, you can see that with an increase of five magnitudes the points representative of these fainter galaxies would fall in a region of the diagram where the curves that define the three models of the universe are widely separated. Thus this method of determining q_0 could tell us with some certainty which type of universe we live in, and we could have a better idea about the future of, among other things, all that surrounds us.

Our long journey through time is finished. We have explored the past of the universe, our past, up to the present time, as well as various possibilities for the future of the world. It has been a difficult journey. Even though astrophysics, geology, and paleobiology have enabled us to observe large segments of the past with our own eyes, much of what we have seen had to be constructed theoretically or by extrapolation from physical laws tested only here on our planet and in the very short span of human history. For these reasons, and because of the relatively few years science has devoted to these investigations, many problems remain to be solved, and many others are just now being defined. In the next few years some of our theories may be replaced by more valid ones, while others may be consolidated by new observational findings, particularly because of the wealth of new data expected from space telescopes; finally, the theories that today leave us perplexed by having more than one solution, like Friedmann's three models for the expansion of the universe, may be shown to have a single valid solution by clear evidence in favor of one or against the others.

Nevertheless, we have made two great discoveries of which we are already certain: that the universe is neither static nor cyclical but has continually changed since it began; and that all changes have occurred in the direction of an increasing complexity, which has led from energy to matter and from matter to thought. Is this the only possible change? Has it occurred and is it occurring by pure chance, or is it ruled by a general law that we cannot yet perceive? What role do living beings play in it? To seek answers to these questions we shall carry out our last explorations, not only rapid excursions in time, but also overviews spanning past, present, and future that will help us to find or sense a less obvious but more profound meaning in all that we have learned.

THE ARROWS OF TIME

As we have seen, and as we experience in every moment of our life, time appears to be asymmetric, in the sense that it seems to flow in only one direction, i.e., from the past to the future. First we must point out that this is not a property of time. It does not flow, as demonstrated by the theory of general relativity, in which time is considered just a dimension. Time asymmetry is an intrinsic property of the universe, where, at least on a macroscopic scale, sequences of events that occur in the same temporal order are solely asymmetric. If we light a match, it will burn, becoming carbon and then ashes; if later on we light another match, it will be transformed in the same way. We shall never see a burned-out match in the ashtray burst into a small flame that becomes brighter and brighter, turns into wood, and then abruptly shuts off leaving a brand-new match. What occurs in our macroscopic world gives us the sensation that it is time that flows and, since it can never turn back, that it flows in only one direction. In reality, it is the processes acting on matter that by occurring always in the same way, as in the example of the burning match, define a one-way, that is, asymmetric, route.

On the other hand, not everything occurs in this manner. In the world of particles, which are in random motion, a distinction between past and future has no meaning. The world of particles is practically symmetric in time.

The fact that all known macroscopic processes occur in only one direction, that is, asymmetrically, is expressed, according to a convention adopted in physics to indicate asymmetries, by an oriented arrow with the point toward the future and the tail toward the past. It is called the *time arrow*.

THE THREE ARROWS
Three time arrows have been defined so far according to the corresponding processes.[50] The simplest is the *cosmological arrow*, which expresses the manner in which the universe varies as a whole. It indicates, for instance, that the universe is expanding and that its average density is continually decreasing. Naturally, this is true now and in the past. As for the future, it will be true only if the universe is flat or open. If the universe is closed, at a certain point it will start to contract and its average density will increase. This will not mean that the time arrow has reversed its direction, but simply that the path from the past to the future is characterized by phenomena that occur in opposite ways. As an example, think of a rubber ball that after falling to the floor bounces back up, retracing its path in the opposite direction.

As for the second arrow, we have discovered it in our journeys into the past when we found that nothing is cyclical and everything evolves, even if we do not know why, from the simpler to the more complex. This arrow has been called the *historical arrow* because inscribed on it is the history of the universe, including our own, in all its details. By showing that the universe changes toward states of greater order and differentiation, the historical arrow tells us that the amount of information contained in the universe increases with each passing moment. When the universe consisted only of energy or particles in random motion, there was very little it could tell us. Currently, the amount of information it contains is very large, much greater than what we can imagine, let alone know. There is much more information today in the brain of a single ant than there was in the entire universe in the era of hadrons or leptons.

Like the previous two, the third arrow points from the past to the future, but it is composed by phenomena that, paradoxically, are the opposite of those defined by the second arrow. Suppose we take half a glass of whiskey and two ice cubes and we put the ice in the whiskey. Little by little the ice will melt, cooling the whiskey, and the melted water will mix with it. When all the ice has melted, the result of this process is a leveling of the system. Consider what we had before: on the one hand, a liquid with certain characteristics, formed of molecules moving freely in all directions and at a certain velocity that determined its temperature; on the other hand, two cubes of a different substance, of a certain shape, at a temperature lower than that of the whiskey, and with the water forming the ice cubes arranged in a crystalline structure according to a geometric and therefore well-ordered pattern. Together, whiskey and ice cubes constituted an orderly system containing a great deal of information. After the ice melts, all that is left are molecules of the two substances mixed together and in random motion, and the only information they give us is about the temperature of one liquid (watered whiskey) instead of the two (ice and whiskey) we had before. In other words, we are left with greater disorder and a smaller amount of information. It is exactly the opposite of what happened with the burning match, which gave heat and light through the complex processes of combustion according to the historical arrow. And yet even in the case of the whiskey and ice cubes we are confronted with an irreversible process because we know of no natural means whereby, on their own, the water molecules could separate from the whiskey and form again ice cubes. Hence this process also defines an arrow of time directed from the past to the future, just like the historical arrow, but unlike the processes that define the latter, this process creates disorder and reduces information. This third one is known as the *thermodynamical arrow*.

This arrow is so named because it was discovered mainly through thermodynamical processes like that of ice melting in whiskey. To understand the nature of such processes better, let us consider two gases (for example, oxygen and nitrogen) at different temperatures and held in two different parts of a container separated by an airtight partition. The different temperatures are due to the fact that the molecules of the hotter gas move randomly at a higher velocity and the molecules of the colder gas move analogously at a lower velocity. If we remove the partition, the molecules of the two gases will mix; as they hit the molecules of the colder gas, those of the hotter gas yield part of their energy to them, and after a while all the molecules will move at the same velocity. In other words, the mixture of the two gases will be at the same temperature. Here again disorder has increased and information has decreased because before we had two distinct groups of molecules at different temperatures and now we have a mixture of two types of molecules at a single temperature.

In fact, this experiment simply shows that heat is always transferred from hotter to colder bodies and never the other way around. This is a way to express the second principle of thermodynamics. Physicists have broadened this principle by introducing a new quantity—*entropy*—which in its more general meaning can be considered a measure of the degree of disorder.

COMPARING THE ARROWS

Given that, as far as we know, heat is always transferred from hotter to colder bodies, the thermodynamical arrow indicates that in the universe entropy must always increase, that is to say, disorder must continuously increase, and the amount of information must continuously decrease.

Let us see what happens when we compare the three fundamental time arrows. The cosmological arrow does not give us any problem when compared with the others. It simply represents the temporal underpinning of an ever expanding universe, or a universe first expanding and then contracting, or an oscillating universe. But the historical and thermodynamical arrows appear to contradict each other. In fact, according to the thermodynamical arrow the universe should have been born very orderly and rich in information and with the passage of time should grow more chaotic and poorer in information, so that at the end all features should be leveled out and no information left. This state is what in the last century was known as "thermal death." Now that the concept of entropy has been extended to nonthermal processes to indicate the degree of disorder, we could simply call it "death" because at that point all is featureless and nothing happens any longer.

But the historical arrow tells us that until now exactly the opposite has occurred. From the Big Bang to the present, complexity, order, and information have always increased and the universe has become richer, more differentiated, and more orderly in its various components. How can these two views be reconciled? It is still an open question, but it has been addressed.

POSSIBLE EXPLANATIONS
The first thing we should bear in mind is that the historical arrow has been traced both on a small scale and in the entire universe, whereas the thermodynamical arrow is based on the second principle of thermodynamics, which, like the law of universal gravitation, has been tested only in our laboratories and extrapolated to cosmic space. We cannot exclude the possibility that there exist processes that counteract the increase in entropy or even cause it to decrease. However, it would be better not to accept this solution unless we actually found such processes. In fact, once certain principles have been established—as few and as reliable as possible—science cannot proceed by accepting or rejecting them according to the different cases.

Many scientists solve the problem by positing the leveling required by the thermodynamical arrow and assume that there is a universal tendency toward maximum confusion and the suppression of all information. In their opinion we see this effect only on a small scale (watering the whiskey and similar processes) because it requires extremely long times. F. Hoyle and J. V. Narlikar write that in the Big Bang cosmology the universe must start with a marked thermodynamical imbalance and eventually must exhaust itself.

D. Layzer, who has devoted many years to the study of the time arrows, has shown that Hoyle and Narlikar's point of view is incorrect, and, in particular, that at the beginning of the universe there need not have been a marked thermodynamical imbalance. In 1976 Layzer proposed a theory of his own according to which the three arrows (cosmological, historical, and thermodynamical) can coexist under two cosmological conditions strictly connected and quite plausible. One, which Layzer calls the *strong cosmological principle*, holds that "no statistical property of the universe defines a preferred position or direction in space." According to this new version of the well-known cosmological principle, even local irregularities must be homogeneous and isotropic like the large-scale parts of the universe. The second postulates that local thermodynamical equilibrium prevailed around the singularity from which the universe originated.

With these assumptions and using sophisticated mathematical tools, Layzer has shown that the universe is not headed toward exhaustion and that the expansion from the singularity has generated macroscopic structure and entropy. If this theory is correct, he concludes that there must be a new type of indeterminacy whereby not even the entire universe contains sufficient information to specify fully its future states. The present would always contain new elements, and the future would never be entirely predictable.

The conclusion that the universe is not entirely predictable accords with that reached by Dyson through a completely different approach. In fact, the growing development of a self-aware life capable of operating changes on an inert matter that should continually diminish in favor of organized matter also produces a sort of general indeterminacy in the universe due to the exercise of willful acts by the living world upon matter. If the theories of Dyson and Layzer are correct, we could conclude that just as it is impossible to know, even theoretically, what happened in the first 10^{-43} second of the life of the universe when everything was extremely simple but unknowable, so we cannot anticipate what will occur in the long run in a universe increasingly more complex and perhaps increasingly affected by intelligent beings whose actions are not dictated solely by physical laws.

IS THE UNIVERSE SPECIAL?

STRANGE EVIDENCE
Following the thermodynamical arrow we have seen that thermal nonequilibrium is a necessary condition for the life of the universe and, as biology confirms, for the existence of our own life. We could not exist

if all matter, including our bodies, were at the same temperature. Moreover, matter could not have evolved to form us and all we see about us (from cars on the road to trees in the woods and stars in the sky) if starting from the remote past there had not been a thermal imbalance prevailing everywhere and manifesting itself in enormous differences in temperature among different bodies. If after its birth the entire universe had remained at the same temperature, our world would have never formed; everything would have remained flat, uniform, and lifeless.

Though perhaps the most important, thermal nonequilibrium is only one of the special conditions that had to occur in order to give rise to the world we know, or even to prevent the universe from falling apart as soon as it formed. Let us consider at least some of the other conditions.

We have already met one of them when we discussed the first instants in the life of the universe. If you recall, when heavy particles formed, more baryons than antibaryons were produced. Hence not all of these particles annihilated each other, turning into radiation, and matter, the fundamental constituent of the present universe, could be born. At first sight it might not appear too strange that the number of baryons should have been slightly greater than that of antibaryons (one more per billion pairs). But on second thought it is rather peculiar. Let us take the first billion pairs of baryons and antibaryons; after they annihilate each other there is one baryon left. Now let us take the second billion pairs. If antibaryons had had the same probability of forming as baryons, at this point we could expect to have an antibaryon left over. And if this had not happened for the first two or ten or hundred billion pairs, it should have happened for the next two or ten or hundred billion pairs. In that case, of course, the two groups of leftover baryons and antibaryons would have annihilated each other. The fact that it did not happen means that baryons had a greater probability of forming than antibaryons. But why it should be so is not clear since the particles are identical except for the symmetry of certain individual characteristics.

This peculiarity is not the only one that occurred at the microscopic level to produce the universe we live in. As Dyson has observed, hydrogen burning in stars and the consequent production of energy occur relatively slowly because the fusion of normal hydrogen nuclei is a process of weak interaction. If it occurred through the strong interaction, as in the case of hydrogen bombs, where the fuel is heavy hydrogen (deuterium), the reaction would be a billion billion times faster, the stars would be very short-lived, and life would not have time to develop on their planets. This is a particular case involving only a small amount of hydrogen and therefore a small percentage of stars and planets. More significantly, if the nuclear interaction between protons had been just 3% stronger, most of the hydrogen in the universe would have turned into helium in the first phases of the Big Bang, prior to the formation of the galaxies, and therefore the stars would not live as long as they do.

There is yet another case in which the stars could not have lived long enough to permit the development of complex and intelligent living beings—if the force of gravity were not such a weak force. As B. Carter has shown, the length of a star's life is determined by the fact that the force of gravity is 10^{40} times weaker than the electromagnetic force. If the ratio of these two forces had been smaller, the life of all stars, including the sun, would have been much shorter, and no form of life would have had time to develop on any planet.

Another strange fact pointed out by P. C. W. Davis is that we would not be here if the current temperature of the universe were 300°K (27°C) rather than 3°K. First of all, 27°C is the normal temperature of the earth, and therefore there would not be the thermal imbalance necessary for life. Furthermore, the galaxies could not have formed, and consequently stars and planets would not have formed. Thus it appears that the initial temperature of the universe and its rate of change were precisely those required to make the universe evolve to its current state.

Perhaps the most impressive of these coincidences is recalled by Dicke. If the initial expansion velocity of the universe had been even a thousandth of 1% lower than it was, it would not have been sufficient to form stars. The universe would have expanded a little, only to contract again quickly. Strange as it may seem, it is as if the mechanism had been precisely tuned so that the universe would expand enough to give rise to stars, planets, and man. On the other hand, stars would not have formed even in the opposite case, that is, if the expansion of the universe had been too rapid, because matter would not have had time to condense.

The existence of so many strange coincidences has also been demonstrated theoretically on the basis of another fundamental characteristic of the universe—isotropy. We saw earlier that the universe is isotropic, which means that it has the same properties in all directions. C. B. Collins and S. W. Hawking have found that this is possible only for a very narrow range of initial conditions. In other words, they have demonstrated that if some very special events had not occurred, the universe would not be isotropic, as observations show it to be. In a sense this conclusion is a synthesis and a theoretical demonstration of what we have just seen: Special circumstances produced this universe in which we live and of which we are part.

As a matter of fact, the problems are not simplified when we turn from the world of atoms, galaxies, and stars to that of living beings. The evolution of life, which we followed only on earth, also raises some intriguing questions. We have often found changes or innovations for which we could not see any necessity in the preexisting world but were indeed "providential" for the course of evolution, just as some cosmic events had been. Let us recall a few of these turning points.

Consider first the formation of prebiotic molecules and the subsequent development of such complex molecules as proteins and nucleic acids, which in turn combined to form the first living cells capable of replicating themselves. Consider next the transition from procaryotic to eucaryotic cells, which were much more complex and above all more versatile; in particular, they had the capacity of aggregating into a new type of organism, the multicellular, which is much richer and more efficient than a single-celled organism.

Obviously multicellular organisms could no longer reproduce by fission. The next simplest form of reproduction would have been parthenogenesis, which even now is used by some animals. Yet what came to predominate was sexual reproduction, a more difficult form to bring about but a more efficient one from the standpoint of evolution, which otherwise would have been much slower. In retrospect its significance is quite evident. But the primitive organisms that initiated sexual reproduction certainly could not have foreseen its advantages and deliberately chosen this approach. And it is hard to see what could have been the "personal" interest that prompted its implementation by so many generations for such a long time.

As we keep moving in the biological world along the historical arrow we find other revolutionary innovations that arose spontaneously. A case in point is the development of the tracheid, the woody cell essential to the existence of high-stemmed plants. Without it, trees, underbrush, humus, primitive terrestrial animal life (strictly connected to vegetable life through the food chain), and finally more advanced animal life, including humans, could not have spread over the face of the earth.

Another very important event was the development of the amniote egg, which permitted the diffusion of animal life in places far from the water. And yet more important, at least from a selfish point of view, was the emergence of the mammals.

But it is useless to continue this list with more or less justified references that we can discover by retracing the biological part of the historical arrow. What is significant is that also in the biological world there were transitions that may be considered strange since at the time they occurred there was no reason why things could not have gone on as they were. But the fact is that changes did occur, and each time life made a leap forward. With the passage of time these leaps became more and more significant and frequent. As a consequence, like the world of galaxies, stars, and planets, the living world became more varied, more complex, and richer in information and possibilities.

THE ANTHROPIC PRINCIPLE

In view of the last considerations, all our journeys in time point to the conclusion that the universe was born and evolved in such a way as to produce an intelligent life that is part of it and is capable of understanding it. Without life, or more precisely without science, which can be developed only by intelligent beings, nature could not have achieved self-awareness. This is the basis for the formulation of the so-called *anthropic principle*. Essentially it says that the fundamental constants that operate in the physical laws governing the universe have their own distinct values and no others so that in the universe there should occur the phenomenon of consciousness.

This principle could apply to any intelligent being, not just man, but since man is the only conscious being we know, it tends to assume an anthropocentric connotation. In other words, it seems to suggest that the universe would have no meaning if we did not exist, which is tantamount to saying that it needs us. From here it is but a small step to say that the universe has in fact been made for man, or at least for thinking beings. It seems to be a return to geocentrism, to the concept of man as the ultimate goal of all things, and by another small step to the concept of man as the king of creation.

But let us not go too far. The anthropic principle does not warrant such conclusions. It merely says that the universe was formed in such a way that one day it would achieve self-awareness. But how could the universe have arranged things in such a rational and balanced way when it was not yet able to understand itself—indeed, when it did not even exist yet? The most logical answer one can give is that the universe is special because, if it were not so and had not been so from the beginning, we would not be here to ask the question. This is one possibility, but at the same time it induces us to reflect on other possible explanations.

CHANCE OR FINALISM?

Let us start again from the biological world. As we have noted, it has developed along the historical arrow, which is the evolutionary arrow from the simpler to the more complex. Darwinism explains biological evolution by the mechanism of chance and many false starts. Let us apply it to the entire universe. If the universe is closed and eventually falls in on itself, converging again toward the singularity, it is possible that through a sort of rebound it will start to expand again, only to fall in again, and so on. According to J. A. Wheeler, an oscillating universe could go through an infinite variety of structures and conditions. The fundamental constants, and perhaps even the laws of nature, could change from one cycle to the next. Thus there would not even be an increase in entropy with each succeeding cycle; each time everything would be reset to zero and would start anew. This oscillating process could be repeated an infinite number of times. If so, after an infinity of cycles in which all kinds of universes were generated, a universe came along that was just right for us, that is, a universe where by chance everything conspired to bring about our existence. In sum, the lucky circumstances needed to produce a living world would have occurred for the same reason that a monkey hitting the keys of a typewriter at random in a sufficiently long time will chance to compose the *Divine Comedy*. If such were the case, the proper interpretation of the anthropic principle is that the universe seems tailor-made for us because otherwise we would not be here.

If the universe is open, however, matters appear in a different light. In this case, there is only one start, and the chance that this one time there should be all the lucky circumstances needed to set in motion the evolutionary process leading to life and man, though still possible, becomes extremely improbable. If there is only one start, then the most natural explanation is that the universe has achieved self-awareness because it was made for that purpose and therefore the necessary conditions were "built in" from the start. From this point of view, the current state of consciousness provided by humankind, and possibly by other thinking beings on other planets, could be just an intermediate step.

The idea that evolution may not be due to chance, or at least not entirely to chance, is gaining some acceptance in biology itself, where people are beginning to doubt that Darwinism is the only explanation for the way the living world evolves. In paleontology, which shows, albeit with many gaps, how animals have changed over hundreds of millions of years, we have the best way to study evolution from factual evidence and over the longest period of time. After devoting many years to this type of research, P. P. Grassé has concluded that "any theory professing to explain evolution must invoke a mechanism other than mutation and chance." If chance and necessity are not the determining factors in biological evolution, or at least not the only ones, then the same may be true for the universe in general, particularly if evolution from the simpler to the more complex occurs only once in an open universe. After all, living matter is nothing but the natural and continuous development of inanimate matter. From the world of galaxies, stars, and planets to the world of bacteria, forests, and man there is no miraculous jump, but rather a continuous process of transformation of matter.

If we exclude chance and the possibility that everything has been accomplished by the universe itself, which has not existed forever, the fact that the universe has been built and designed toward the achievement of self-awareness may lead one to feel again the need for a Creator existing outside of matter, space, and time.

If so, He is the author of the design that is being carried out for a purpose; it was He who created space and time and provided the initial impulse; and it is to Him that the universe as a whole is continually tending through an ever increasing complexity and spiritualization of matter, even though it might require an infinite time to reach Him.

Thus discovering if we live in a closed (oscillating) universe or an open one could be very relevant to the solution, admittedly personal and nonscientific, of the greatest metaphysical question of all times, namely, the existence of God and how His presence is felt throughout the universe.

THE ROLE OF THE LIVING WORLD

HUMANKIND

For centuries, for millennia, we have asked ourselves where we are, where we came from, where we are going, and what role we play in the world, and we have looked for answers in superstition, religion, philosophy, and science. We have known the answer to the first question for some years now, thanks to astronomy. Science has also told us where we came from and is helping us to glimpse an answer to the other two questions. Science has not told us yet where we are heading, and almost certainly it never will because the answer depends in part on man himself. But science is beginning to answer something even more important, that is, the last question: What does man represent in the general economy of the universe?

As far as we know, we are the highest development yet achieved by primordial hydrogen, or better, by the primordial energy that turned into matter and radiation and then progressed to more complex forms. From energy came particles, from particles came atoms, from atoms came simple molecules and then more complex molecules, while the whole was condensing into immense clouds that generated a myriad of stars and an even greater number of cold, solid bodies, where the evolution of matter progressed to the first cells, blossoming into life. Life then continued to evolve into more and more complex forms under the influence of the environment, which, in turn, was increasingly modified by living beings. Plants and animals developed richer and richer structures until one of them acquired thought. As we have seen, it is through thought that the universe has become aware of itself.

But this is not the only consequence of the advent of intelligence. Thinking beings, who give greater value to the universe by understanding it, also have another faculty, that is, the ability to act. Every work is a creative act of human thought. It is performed by an artist, a philosopher, a mathematician, any one of us. It can be more or less intense depending on one's intelligence and fantasy, but it always has consequences that reverberate throughout the surrounding world. By the ability to act, matter has taken a further step. It has surpassed itself by creating something that it could have never produced without going through life: one of Phidias's statues, Shakespeare's plays, or the geometry of hyperspace.

The two routes have almost always been traveled separately by individual people; the scientist who tried to understand the universe would ignore the artist who created, and vice versa. But now we know that each of our thoughts, each scientific or artistic work, contributes to the advancement of the spiritual part of the universe. If we appreciate this responsibility, we feel that we have discovered what our task is and what is the best way to spend our lives.

Naturally it can only be stated in general terms, but even with this limitation the answer seems quite clear. In every work of art there are ephemeral values tied to the society and time in which it is created, as well as values that are universal and eternal (using these words here in a literary rather than a literal sense). It is through these values that our works speak to people of all times and all places.

The life of each of us is also a work of art, and each of us is the artist who creates it by working on himself. And also in this creation that is our lives there are transient values, valid only for the time and place in which we live, and lasting values, which will emerge everywhere and always. And the more we understand the world in space and time, the better we see the distinction. On the one hand, there are the thousand things of everyday life—some trivial, some necessary but ephemeral—which at the time may seem important and urgent, and sometimes are, but do not endure beyond the moment; on the other hand, there are the things that have a lasting value from the cosmic point of view and are essentially embodied in two acts: discovery and creation.

Nature has shaped us in such a way that we are forced to act in the first way, for example, in order to tend to our bodies and in general to the requirements of individual and social life, starting with the acts necessary to survival. But it has also given us the means and the stimuli to realize the second. We can enhance our work of art one way or the other by making of our whole lives either a modest copy of what is fashionable or unique creations rich in information of lasting value that enrich the world around us.

UNIVERSAL LIFE

Now that we know a little more about our place in the universe, let us return once again to the question of what happened elsewhere. We do not know, but the uniformity of behavior of the physical evolution of celestial bodies, for which we have found so much astronomical evidence, leads us to believe that something similar to what happened on the earth has occurred, is occurring, or will occur in many parts of the universe. Perhaps life has appeared in different guises, with different ways of feeling and thinking. Somewhere in the cosmos there must be worlds going through the same evolutionary phases the earth experienced millions or billions of years ago. But there may also be worlds where biological evolution started earlier. And if it occurs elsewhere at the same accelerating pace we observe on the earth, a great many of the worlds where the life process started a billion years earlier than on ours would now be in a different and much more advanced phase of development. A billion years before the birth of our planet the universe certainly was such as to permit the formation of planetary systems. Hence it is more than likely that in the universe there are forms of life far more advanced than our own.

Still, we have no proof of all this. If we manage to preserve and advance our civilization, it is possible that one day we shall meet superior beings, share their knowledge, thus rising quickly to their level, and then continue together on the path of learning and creativity. But for the time being, as far as we know, human thought is the highest point reached by matter in the cosmos.

Should this be the case, the consciousness achieved by the universe would be confined to a small planet that is as a drop of water in the ocean. It may seem absurd that a universe of a vastness and a richness beyond the comprehension of the human mind should become aware of itself through such a mind. It becomes almost tragicomic when you think that for the majority of people the universe is no bigger than their town or, at most, a little larger than the planet earth. This calls to mind Andersen's tale of the duck that explained to her ducklings that the world did not end at their nest but went as far as the miller's house and even a little farther.

There is no cause for regrets, however. The anthropic principle would be just as valid if it were to apply to a single thinking being. But it seems impossible that it should be so restricted, and a moment's reflection shows that it need not be for two reasons that may coexist.

First of all, the fact that most people do not know the universe in which they live is only a temporary condition. The number of those who confront the issue, and hence become a part of cosmic self-awareness, is constantly rising. If it continues this way, the day may not be too far off when all of humankind will be as knowledgeable as the scientists themselves, at least with regard to the essential facts. Furthermore, the spread of knowledge may not be confined to the earth. If man can extend his ecological niche into space, the places where the universe knows itself and is transformed by a process born of intelligence will be ever more numerous and distant.

Even with space travel and space colonies, however, the expansion of terrestrial life would involve only an infinitesimal part of the cosmos. But there may be another way for the universe to reach the high degree of self-awareness and creativity provided by highly intelligent and sensitive forms of life, namely, if such beings were to develop on an enormous number of planets, whence knowledge and creativity could spread farther and farther. These two events—the birth of intelligent life in so many places and its development in and from each of them—would add up and multiply until the universe would become like a single body. Within it, the separate and different societies could be like the cells of an organism that function independently but at the same time are part of a living being with its own individuality and a new kind of thought—a thought that the cells do not share but can only exist because of their contribution.

In sum, the formation and aggregation of so many conscious and creative units could make the universe into a whole that is more than the sum of its parts. This is an idea that our mind can accept because it would be analogous to what happens in our body through its constituent cells. It is more difficult to conceive of the universe as a "thinking" entity, particularly because we could not imagine its thought or how such a thought could manifest itself. On the other hand, even though the cells of a living organism carry a great deal of information, in no way could one of Beethoven's red blood cells have perceived that the organism to which it belonged was composing the Ninth Symphony. To the red blood cell this act had absolutely no meaning, even though it was contributing to it, along with millions of other cells, by oxygenating the blood of the composer while he was writing. In sum, the universe might not only evolve in its constituent parts, but through many partial acts of learning and creativity it might become an organic entity with an activity bearing the same relation to ours as composing the Ninth Symphony was to the life and function of one of Beethoven's red blood cells.

Here we stop and look back, not at the past of the universe, but at the last pages of this book. Our mind has worked a great deal, but we are beginning to think that fantasy has run on even more. Perhaps so. Yet there is a scientific basis for the concept of the universe as an organic whole. It emerged nearly a century ago from Mach's criticism of the mechanistic view and, in particular, from what Einstein called "Mach's principle." In its simplest form, Mach's principle states that the behavior of one part of the cosmos is determined by all its other parts. This principle recurs in the cosmological theory of the steady-state universe (which we have not discussed because it has been supplanted by the Big Bang theory) and again in a more recent theory formulated by Hoyle and Narlikar. Their view of the universe as an organic whole is worth mentioning, particularly because the "special" characteristics of our universe seem to call for such a concept even in the case of the Big Bang.

The classical universe—the universe in which planetary and stellar orbits were so harmoniously distributed—was ruled by a mechanism whereby the various parts influenced one another, but not in a determining manner. To give an example, a planet or a galaxy could disappear without any other consequence than a slight rearrangement of the whole. Such a universe could be taken apart and reassembled like an automobile, whose various parts are related in a purely "external" way, so that after being reassembled the automobile would work as well as before.

But according to Hoyle and Narlikar, the properties, and even the existence, of any part of the universe is conditioned by the other parts. According to these cosmologists, for example, the mass of the electron depends on the number of another type of particle, the μ meson, and the mass of the μ meson depends on the number of electrons. Such a universe could not be taken apart without ruining everything, just as we cannot take apart a living organism without destroying it as such.

As we explored the universe through time, rather than finding a confirmation of the mechanistic view, we have often had the feeling that something more complex was at work among its parts and on the whole, something more akin to intelligence than to stone.

Thus it is possible that we have not forsaken reality for the boundless realm of science fiction and that our intuition and the theories mentioned above are showing us the right way. After all, this extraordinary universe that in time becomes so rich, so complex, and so harmonious as to appear almost like a single entity is not very different from Dyson's view of it. If such a universe could exist from the mathematical and physical point of view, as Dyson has shown, why could it not become a reality?

ETERNAL EVOLUTION

In the last century, with an imagery somewhat romantic and rhetorical, the course of history was portrayed as a relay race with each athlete representing a generation, and with the torch of civilization and progress passing from one generation to the next and burning ever more brightly. Although there is some truth in this image, it is much too limited.

In the first place, we cannot imagine the torchbearer as an athlete with a human body of classical beauty. The torch passing from one generation to the next was already a small flame when it arrived at the first unicellular beings. In turn, they handed it to more complex beings, the primitive fish, from which it passed to the fish that left the water, then to amphibians, reptiles, and small mouselike mammals, from which it passed to more advanced and intelligent mammals and finally to man. The torchbearer has changed repeatedly in the past and will continue to change. When man can no longer carry it, it will pass to other species, either existing species that are now considered "inferior" or new species into which man will evolve.

Furthermore, the nineteenth-century image is too biased in another sense. Today we no longer believe in a single civilization that advances and expands to bring light to the savage people around it. Rather, we see many different cultures, each deriving its character from the environment in which it develops, from the past whence it comes, and from the interactions of its various aspects. When we say "man" we do not mean only man and woman but humankind. Humankind is the sum of all cultures, and as these change and interact it becomes more complex but also richer because in the process each culture, each individual, loses something of itself but gains from the others. Thus there are cultures that have contributed more, but there are no cultures that have contributed nothing. There are no superior or inferior cultures, only different cultures.

In its essence, this view could be valid on a cosmic scale. The idea that we may not be the only torchbearers in the universe emerged in the last century (and sporadically even earlier). Today we have many more reasons to believe that what is occurring on the earth is also taking place on countless other worlds. The unitarian evolution of the universe leads us to believe that there are not many independent relay races, one on our planet and the others on as many other worlds, already run or now running or to be run by various types of humanoids ignorant of each other's existence.

We can still use the analogy of the relay race, but we must adapt it to our current view of reality. In this more comprehensive view, the torchbearer is the stuff that evolves into living matter and proceeds toward goals that we cannot perceive. The torch that is carried and handed down is the deed, at first performed unconsciously, and then, with the advent of intelligence, in a conscious and purposeful way. The starting gun was the cosmic explosion. It was then that the first runner, energy, started off; since then the relay race has gone on in the entire universe with many changeovers, at first very rapidly, then slowly, then faster and faster, perhaps to end with infinite slowness.

We now know that there was an epoch, that of hydrogen and helium, in which there was no trace of life anywhere in the universe. It was followed by a long period, dark but active (a sort of cosmic Middle Ages), in which preparations were being made for a new epoch that would see the formation of planets and the emergence of life on earth. These events, coupled with the diffusion of prebiotic molecules in space, lead us to believe that what has taken place on the earth in the past few billion years almost certainly has occurred, is occurring, or will occur on innumerable other worlds.

The nineteenth-century image of evolution has been crushed, shattered, and extended well beyond the confines of the earth. We can no longer view it as the advancement and expansion of our living world, but as an action that involves the entire substance of the universe in a continuous global process of increasing complexity that elevates matter to ever more refined states of conscious spirituality.

APPENDIX 1   STRUCTURE AND EVOLUTION OF A RED GIANT

In chapter 2 we explored the interior of a sunlike star, but we could not linger on the structural details of a red giant without digressing from our general survey of stellar evolution. However, it is worth exploring the interior of a red giant and some of its evolutionary phases, even if it is a journey that we can make only in thought on the basis of the results of theoretical research. Guided by I. Iben Jr., who in 1971 calculated all the pertinent data, we shall discuss the case of a star of $0.7M_\odot$, a little lighter and smaller than the sun.

Figures 91 and 92, both to scale, illustrate the interior of the star after it has finished burning hydrogen in its core and has evolved into a red giant. Figure 92 shows a greatly enlarged section of the core (the black dot in the center of figure 91). Due to the burning of hydrogen in a shell 6,000 km thick (figure 92), the star, which initially had a radius smaller than that of the sun, has become 50 times larger, while its luminosity has risen to almost 1,000 times that of the sun. Its surface temperature has dropped from 6,200°K—the temperature it had while on the main sequence—to 4,400°K; hence the star has turned from yellow to red. The strangest thing is that although the star has swelled into a giant, most of its mass and energy production is concentrated in a sphere just 5 times the size of the earth, represented, to scale, by the black dot in figure 91 and greatly enlarged in figure 92. The helium core is degenerate, but the new helium being formed by the burning of hydrogen in the overlying shell, shown in white, keeps on increasing its mass and causes a contraction that develops heat to the point of kindling the helium in the interior. In figure 92 this phase has not been reached yet because helium burns at 100 million degrees while, as shown by the figure, the maximum temperature, which does not correspond to the central one, is still 72 million degrees.

When the helium in the core reaches a temperature of 100 million degrees it starts burning with the "flash," which according to the latest studies occurs more than once during a very short period of time (not over ten years). During this phase the surface temperature of the star rises rapidly; its luminosity, however, declines and then stabilizes because its radius decreases until it is only 4 times that of the sun. From red the star turns white.

This phase is illustrated in figures 93 and 94, the latter again showing an enlarged section of the core. The star's luminosity is now 43 times that of the sun, and its surface temperature is 7,250°K. Figure 93

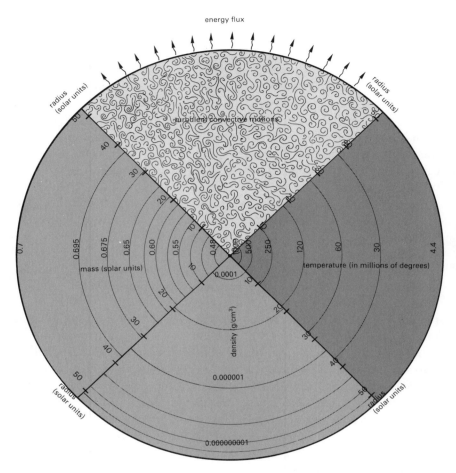

Figure 91   Section of a red giant showing some aspects and physical character-istics of its interior. The star has been divided into four quadrants for the sake of clarity. For a realistic view, one must imagine all these characteristics distrib-uted within a globe of gas that merges with space beyond the outer circumfer-ence. [From *Le Science* (October 1970)]

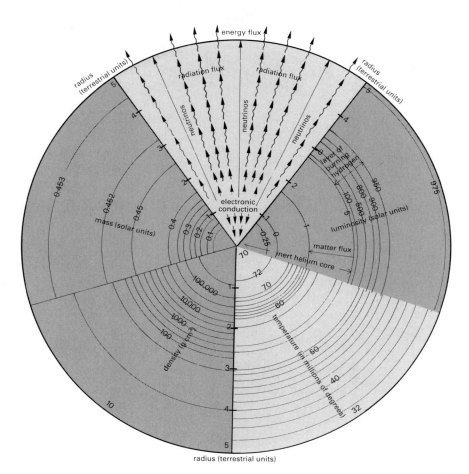

Figure 92   Enlarged section of the core of the red giant—the black dot at the center of the previous figure. Note that the radius is just five times that of the earth. Here, too, one must visualize all the characteristics of the different sections distributed in concentric layers within the sphere representing the core. [From *Le Scienze* (October 1970)]

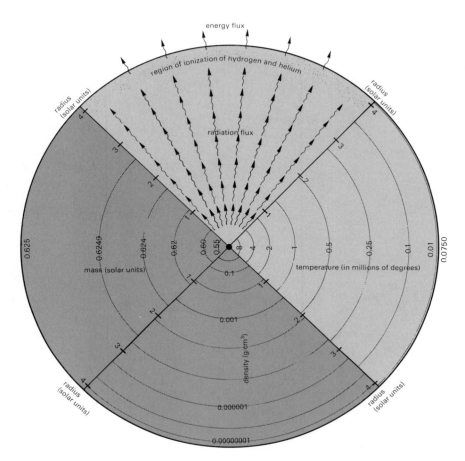

Figure 93   The red giant has become nearly white and has moved in the H-R diagram to the horizontal branch. Later, however, it will expand again into a red giant. The black dot at the center represents the volume within which the energy of the star is generated, and is shown enlarged in figure 94. For this figure and the next, keep in mind what was said earlier concerning the division into quadrants. [From *Le Scienze* (October 1970)]

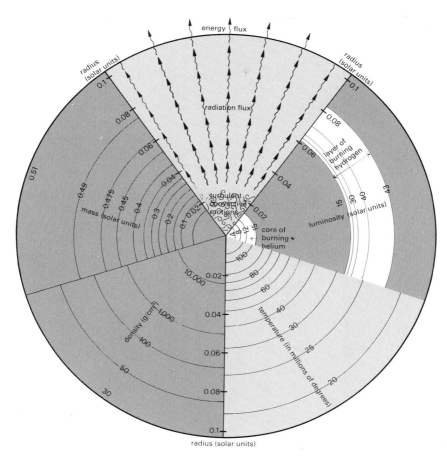

Figure 94   Enlarged section of the core of a star on the horizontal branch, represented in the previous figure by the black dot at the center. Although its radius is only a tenth that of the sun, the core contains more than 80% of the star's total mass. Deep in its interior helium burns into carbon at a rate 100 times higher than the rate at which hydrogen burned into helium when the star was beginning its life on the main sequence. [From *Le Scienze* (October 1970)]

shows that the mass of the star has diminished slightly. This can be explained by the fact that the star must have lost mass during the red-giant phase. Though still very small, the core is now more than twice as large as it was during the red-giant phase. At the center there is a region (shown in white) where helium burns into carbon, separated from the shell of burning hydrogen by a layer of helium that does not burn (dark gray region) because it is below the kindling temperature, as the figure itself shows. The helium burning in the central region forms a new core of carbon and oxygen, and above it, once helium is exhausted in the center, a shell will form in which the helium still inert in figure 94 will start burning, as occurred earlier with the hydrogen shell. The star will expand again, once more appearing as a red giant.

APPENDIX 2   SOME REMARKS ON CHANDRASEKHAR'S LIMIT AND ON THE EVOLUTIONARY PATHS OF STARS

In 1935 S. Chandrasekhar found that, due to its structure, a white dwarf cannot have a mass greater than a certain value, which according to the latest calculations is $1.44\ M_\odot$. However, as L. Mestel first demonstrated in 1965, this upper limit applies only to a star that has spherical symmetry and does not rotate. While the first condition may apply to the majority of stars, the second almost certainly applies to none because, as far as we know, all stars rotate. In 1966 J. P. Ostriker, P. Bodenheimer, and D. Lynden-Bell found, still theoretically, that there could be stars with differential rotation, of the type the sun has, for which there would be no upper limit at all.

In any case, the value of the critical mass, that is, the value above which a white dwarf cannot form, also increases if there is a magnetic field. We have known for a long time that many stars have appreciable magnetic fields, and recently this has been found to be the case for a number of white dwarfs as well. For all these reasons the value of the critical mass is no longer so easy to define. The upper limit of $1.44M_\odot$ calculated for hypothetical stars certainly cannot be the only one; there must be another one for real stars that rotate, have a magnetic field, are not spherical, and so forth. In this case it would be much higher— say, 3 or 4 solar masses and perhaps more.

I should also point out that the evolutionary paths described in chapter 2 were calculated more than ten years ago for ideal stars, that is, stars with spherical symmetry, nonrotating, and devoid of a magnetic field. Normally this simplification does not have serious consequences because most stars are nearly spherical and have a rather weak general magnetic field. The effect of rotation is much more significant, but only in the case of stars of the first spectral types with masses in excess of $1.5M_\odot$. In 1978 the Soviet astronomer G. S. Bisnovatyi-Kogan and his coworkers calculated new models of protostars and their evolutionary paths on the H-R diagram taking rotation into account. In particular, they considered stars of 0.5, 1, 2, and 10 solar masses. With increasing rotational velocity the protostar takes a longer and longer time to reach the main sequence, and if we bring the rotational velocity to a maximum this period of time becomes even twice as long. In the more massive stars rotation also causes a change in internal structure; for example, the central radiative region forms very early, during the phase of gravitational contraction.

In practice, these theoretical studies, some of which are still in progress, show that the effects that had been disregarded, like the effect of rotation, falsify the results described in chapter 2, especially with regard to the more massive stars. The models I have used, calculated before the 1970s, represent simplified cases that do not correspond to conditions in actual stars but in most cases approximate them quite well. The latest research, instead, takes into account effects previously disregarded that make calculated stars more like the real ones. But though more accurate, the new results have not changed the essence of the previous ones.

APPENDIX 3   SOME REMARKS ON THE CLASSIFICATION OF DINOSAURS

Since in the text I have tried to reduce the terminology to a minimum, I thought it might be useful to write a short summary of the main definitions and characteristics.

The word "dinosaur" was coined in 1842 by R. Owen from two Greek words, *deinos* (terrible) and *sauros* (lizard), to convey the idea of animals of gigantic proportions. As it turned out later on, not all of them were large; indeed, many were quite small.

Dinosaurs are divided into two distinct orders, Saurischia and Ornithischia. These names refer to the difference in their pelvic structures, which is the main distinguishing characteristic between the two groups. As you may know, in all four-limbed vertebrates the hind limbs are connected to the spine by a set of bones called the pelvic girdle. In reptiles the pelvic girdle consists of three bones: ilium, ischium, and pubis. In the saurischians the ischium and pubis were elongated and divergent; the pubis extended downward toward the front, while the ischium ran down the back toward the tail. In the ornithischians pubis and ischium extended backward, were thinner, and were partly welded.

The dinosaurs belong to the class *Reptilia*, subclass *Archosauria*, although some authors prefer to assign them to a separate class (*Dinosauria*). Table 8 summarizes their main characteristics by order and suborder.

Table 8   Orders and suborders that encompass the dinosaurs (class, *Reptilia;* subclass, *Archosauria*) and principal characteristics

Order	Suborder	Characteristics
Saurischia (pelvic girdle typical of reptiles)	Theropods (with predator feet)	Carnivorous, almost all bipedal, can be better distinguished in Coelurosaurs (smaller carnivores), Carnosaurs (large carnivores), and Dromeosaurs (last carnivorous dinosaurs)
	Prosauropods	Ancestral forms of Sauropods living in the Triassic; more primitive and smaller; both bipedal and quadrupedal
	Sauropods (with lizard feet)	Herbivorous; the largest animals that lived on land; all quadrupedal
Ornithischia (birdlike pelvic girdle; all herbivorous)	Ornithopods	Bipeds
	Stegosaurs	Quadrupeds
	Ankylosaurs	Quadrupeds
	Ceratopsians	Quadrupeds

NOTES

CHAPTER 1

1. [A light-year is an astronomical unit of length equal to the distance that light travels in one year in a vacuum, or about 5,878,000,000,000 miles.—Trans.]

2. The observer is assumed to be located in the temperate zone of the Northern Hemisphere—for example, in Italy. The reasoning is also valid for the Southern Hemisphere, but the reference points must be reversed.

3. At present the age of the universe is very much a matter of controversy because it depends on the Hubble constant (H), for which current measurements give conflicting values. I have adopted the age of 15 billion years following the 18th General Assembly of the International Astronomical Union (August 1982), during which various participants discussed the latest data on the Hubble constant and presented evidence that the oldest stars were formed at least 15 billion years ago.

CHAPTER 2

4. For an introductory description of the formation of stars and their birthplaces see *Beyond the Moon*, chapter 5.

5. See *Beyond the Moon*, chapter 1, "In the Sun's Interior" and "The Crucible of the Elements."

6. Chapter 4 of *Monsters in the Sky* is entirely devoted to the history and interpretation of this extraordinary and still mysterious celestial object.

7. Although it is convenient to use the word "burning," it should be kept in mind that the thermonuclear reactions in a star's interior are a very different process from the combustion of logs in the fireplace.

8. See *Beyond the Moon*, chapter 4, "Pulsating White Stars."

9. The formula that permits us to calculate the age of a cluster, given by Sandage and Schwarzschild, is

$$t = 1.1 \frac{10^{10} M}{L},$$

where M and L are the mass and luminosity in terms of the sun and t is expressed in years.

10. See *Beyond the Moon*, chapter 3, "Exceptional Stars: White Dwarfs."

11. I have adopted the value $0.2 M_\odot$ on the basis of R. F. Webbink's calculations.

12. The evolutionary paths were calculated by I. Iben Jr. by assuming that the chemical composition of the original material was 70% hydrogen, 27% helium, and 3% other elements for stars of $30 M_\odot$; for all the other stars he assumed a chemical composition of 708‰ hydrogen, 272‰, helium, and 20‰ other elements.

13. See *Monsters in the Sky*, chapter 3, and in particular "The Supernovae."

14. See *Monsters in the Sky*, chapter 7, pp. 319–322.

15. For the history of their discovery see *Beyond the Moon*, chapter 3, "Exceptional Stars: White Dwarfs."

16. Not all stars can become white dwarfs. If the mass exceeds a certain value known as the Chandrasekhar limit the white dwarf cannot from (see appendix 2).

17. For more information concerning planetary nebulae see *Monsters in the Sky*, chapter 3, "Gaseous Nebulae."

18. Typical examples are FG Sge, HD 50138, HD 45677, MWC 342, MWC 349, MWC 300, M 1–92, M 2–9, V 605 Aq1, V 1016 Cyg, and HM Sge.

19. See *Beyond the Moon*, chapter 4, "The Supernovae," and *Monsters in the Sky*, chapter 3 and chapter 4, "Type V Supernovae."

20. Chapter 5 of *Monsters in the Sky* is entirely devoted to this subject.

21. This spectroscopic effect enables us to discover expanding stellar envelopes. For more information see *Monsters in the Sky*, chapter 4, "Slow Novae."

22. See *Beyond the Moon*, chapter 3, "Double Stars," and chapter 4, "Eclipses of Stars."

23. According to Gorbatsky's theory, stars of the W Ursae Majoris type would be born from a single star that is in an advanced red-giant phase, the hydrogen in its core having already turned into helium. At this point the dense helium core, all at the same temperature and in very rapid rotation, would break up into two separate stars. This theory has some weak points. W Ursae Majoris stars have been found in young associations and presumably are themselves no older than some tens of millions of years. Hence they would not have had enough time to form a helium core, which takes billions of years. This process would be much more rapid in the case of very massive stars, but observations show that W Ursae Majoris stars have masses of only about 1–$2M_\odot$.

24. The previous phase of the primary star may be different, but I have taken the most typical case.

25. See *Monsters in the Sky*, chapter 3, "Runaway Stars."

CHAPTER 3

26. By sphere of action of a planet we mean the sphere centered on the planet within which its gravitational attraction on another body (asteroid, comet, etc.) is stronger than that of the sun. See *Monsters in the Sky*, chapter 1, "The Real Orbits: Orbital Perturbations," footnote on p. 18.

27. See *Monsters in the Sky*, chapter 1, "End and Origin: Origin."

28. By angular momentum of a system (e.g., a body rotating about an axis) we mean the product of the mass and the linear velocity by the distance from the rotation axis. If the system is isolated, the angular momentum does not vary with time.

29. The accretion theory does not entail that the mass or volume of today's moon should correspond exactly to the sum of the masses or volumes of the protomoons. For example, there could have been a system of 4 protomoons, each half the mass of the moon, or a system of 6 protomoons, each a tenth the mass of the moon.

CHAPTER 4

30. I have excluded viruses. Since they reproduce only in living cells and lack an independent metabolism, viruses may be regarded as being at the boundary between inanimate matter and living organisms.

31. Symposium "Life in the Precambrian" held in Leicester, England, 10–12 April 1980.

32. See appendix 3.

33. Some paleontologists disagree with the view of a sudden worldwide catastrophe and lean toward a process of gradual extinction or transformation.

34. More about this process when we discuss the end of the world.

35. For an introduction to this subject see *Monsters in the Sky*, chapter 3, "Super-Explosions: The End of the Dinosaurs."

36. The names "artiodactyl" and "perissodactyl" indicate, respectively, that in the former the legs rest on the third and fourth toes, and in the latter that the limbs have only three toes and sometimes the third toe can constitute the whole foot, in the form of a hoof, as in the horse.

CHAPTER 5

37. This would also provide supporting evidence for the theory of a cosmic explosion as the cause of the extinction at the Cretaceous-Tertiary boundary.

38. I have derived this example from the correspondence between the Latin scholar C. Trabalza and the philosopher B. Croce. At that time the mail traveled by train from Perugia to Naples via Rome, for a total of 456 km.

39. Many research groups have envisaged the future of humankind on the basis of present trends. Listed in the bibliography are a few pertinent books and articles, which in turn will furnish the reader with additional bibliography on the subject.

CHAPTER 6

40. See chapter 2 of this book and chapter 5 of *Beyond the Moon.*

41. See *Monsters in the Sky,* chapter 1, "Origin."

42. This value is somewhat optimistic. If we eliminate stars of spectral types earlier than F5 (see chapter 3), we can obtain it only by assuming that there are mechanisms different from the process of accretion theorized for single stars. No such mechanisms are known, but they may exist, particularly in binary systems. On this matter see T. A. Hepenheimer, *Astron. Astrophys.* 65 (1978):421.

CHAPTER 7

43. For a more detailed description see *Beyond the Moon,* chapter 6, "Other Groups: Other Stars" and "A Look at the Cosmos."

44. In astronomy all the elements other than hyrogen and helium are called "metals." Though incorrect, this terminology is very convenient in that we can denote by it all the elements heavier than hydrogen and helium, which are practically the elements built by the stars.

45. See *Beyond the Moon,* chapter 7.

CHAPTER 8

46. This word does not have a Greek etymology. It was taken by Gell-Mann from James Joyce's *Finnegan's Wake,* a work in which the author delighted in inventing or changing words to suggest new meanings. The phrase is "three quarks for Musther Mark," which suggests "three quarts (of beer, perhaps) for Mister Mark." And the quarks introduced by Gell-Mann were in fact three. Although new in the English language, quark is actually a German word that means "curd" and figuratively "trifle."

CHAPTER 9

47. See *Beyond the Moon,* chapter 10, "The Galaxies Move" and "A Cosmic View."

48. See *Monsters in the Sky,* chapter 5, "Types of Black Holes."

49. See *Beyond the Moon,* chapter 10, "A Cosmic View."

CHAPTER 10

50. Indeed, there is a fourth arrow, the *microscopic arrow,* which is defined by the decay of the K^0 meson. We do not consider it here because it has no direct cosmological implications.

BIBLIOGRAPHY

The following bibliography is ordered by subject matter (and thus by chapter) and by chronology (the most recent works listed first). The bibliography contains (1) original papers and (2) popular books and articles through which, among other things, one can obtain fuller information as to (1). Concerning (2) precedence has been given to the writings of the same scientists who made, or contributed to, the discoveries discussed in them. Essential reading—as also shown by the frequent references in the notes—are two of my previous books: *Beyond the Moon* (Cambridge, MA: 1978) and *Monsters in the Sky* (Cambridge, MA: 1980).

CHAPTER 2   ORIGIN, LIFE, AND END OF THE STARS

Chaisson, E. J. "Le nebulose gassose." In *Le Scienze* (February 1979).

Meadows, A. J. *Stellar Evolution*. Oxford (1978).

Morrison, N. D., and Morrison, D. "Mass Loss and the Evolution of Massive Stars." In *Mercury* (July–August 1978).

Shklovskii, I. S. *Stars. Their Birth, Life, and Death*. San Francisco (1978).

Dickman, R. L. "I globuli di Bok." In *Le Scienze* (October 1977).

Herbig, G. H. "Eruptive Phenomena in Early Stellar Evolution." In *Astrophys. J.* 217 (1977):693.

Terzian, Y. "Recent Findings about Planetary Nebulae." In *Sky and Telescope* 54 (1977):459.

Webbink, R. F. "Evolution of Helium White Dwarfs in Close Binaries." In *Mont. Not. R.A.S.* 171 (1975):555.

Novotny, E. *Introduction to Stellar Atmospheres and Interiors*. New York (1973).

Warner, B. "More High-Speed Photometry of Cataclysmic Variables." In *Sky and Telescope* 46 (1973):298.

Warner, B., and Nather, R. E. "High-Speed Photometry of Cataclysmic Variables." In *Sky and Telescope* 43 (1972):82.

Giannone, P., and Giannuzzi, M. A. "L'evoluzione delle stelle. Le stelle doppie." In *Coelum* 37 (1969):256 and 38 (1970):10.

Harp, H. C. "The Hertzsprung-Russell Diagram." In *Handbuch der Physik* 51 (1958):75.

CHAPTER 3   FORMATION OF THE PLANETARY SYSTEMS

Safronov, V. S., and Ruskol, Y. L. "The Origin and Initial Temperature of Jupiter and Saturn." In *Icarus* (1981).

Coradini, A., Federico, C., and Magni, G. "Origine dei pianeti." In *Giornale di Astronomia* 6 (1980):175.

La planetologia. Roma (1978).

Tai Wen-sai, Chen Dao-han. "Critical Review of Theories on the Origin of the Solar System." In *Chinese Astronomy* 1 (1977):165.

Ruskol, Ye. L. *Origin of the Moon.* Moscow (1975, in Russian). English translation: NASA, TT F-16, 623.

Goldreich, P., and Ward, W. R. "The Formation of Planetesimals." In *Astrophys. J.* 183 (1973):1051.

Nieto, M. M. *The Titius-Bode Law of Planetary Distances: Its History and Theory.* Oxford (1972).

Safronov, V. S. *Evolution of the Protoplanetary Cloud and Formation of the Earth and the Planets.* Moscow (in Russian, 1969). English translation: NASA, TTF, 677.

CHAPTER 4   THE TRANSFORMATIONS OF THE EARTH

Eigen, M., Gardiner, W., Schuster, P., and Winkler-Oswatitsch, M. "L'origine dell'informazione genetica." In *Le Scienze* (June 1981).

Goldsmith, D., and Owen, T. *The Search for Life in the Universe.* Menlo Park (1980).

Sergin, V. Ya. "Origin and Mechanism of Large-Scale Climatic Oscillations." In *Science* 209 (1980):1477.

Desmond, A. J. *L'enigma dei dinosauri.* Rome (1979).

Grassé, P. P. *L'evoluzione del vivente.* Milan (1979).

Russell, D. A. "The Enigma of the Extinction of the Dinosaurs." In *Ann. Rev. Earth Planet. Sci.* 7 (1979):163.

York, D. *Il pianeta Terra.* Turin (1979).

Ageno, M. *La comparsa della vita sulla Terra e altrove.* Turin (1978).

Colbert, E. H. *Animali e continenti alla deriva.* Milan (1978).

Frey, H. "Origin of the Earth's Ocean Basins." In *Icarus* 32 (1977):235.

Irving, E. "Drift of the Major Continental Blocks since the Devonian." In *Nature* 270 (1977):304.

K-TEC group. "Cretaceous-Tertiary Extinctions and Possible Terrestrial and Extraterrestrial Causes." In *Syllogeus* 12 (1977).

Wilson, E. O. *La vita sulla Terra.* Bologna (1977).

Articles on the explanation of the glacial epochs published in *Nature* 260, 261, 262 (1976).

Il pianeta dell'uomo. Milan (1975).

La riscoperta della Terra. Milan (1975).

CHAPTER 5 THE END OF THE WORLD

THE END OF THE EARTH
Grieve, R. A. F., and Robertson, P. B. "The Terrestrial Cratering Record. I and II." In *Icarus* 38 (1979):212, 230.

Wetherill, G. W. "Gli oggetti Apollo." In *Le Scienze* (May 1979).

Combes, M. A., and Meeus, J. "Nouvelles des Earth-grazers—2." In *L'Astronomie* 91 (1977):11.

Jacchia, L. G. "A Meterorite That Missed the Earth." In *Sky and Telescope* 48 (1974):4.

Meeus, J., and Combes, M. A. "Les Earth-grazers (ou EGA), des petits astres qui frolent la Terre." In *L'Astronomie* 88 (1974):194.

THE END OF MAN
Peccei, A. *Cento pagine per l'avvenire.* Milan (1981).

Cullen, C. "Was There a Maunder Minimum?" In *Nature* 283 (1980):426.

Caglioti, L. *I due volti della chimica.* Milan (1979).

Rood, R. T., Sarizin, C. L., Zeller, E. J., and Parker, B. C. "X- or γ-Rays from Supernovae in Glacial Ice." In *Nature* 282 (1979):701.

Eddy, J. A. "Il caso delle macchie solari mancanti." In *Le Scienze* (September 1977).

Tucker, W. H. "The Effect of a Nearby Supernova Explosion on the Cretaceous-Tertiary Environment." In *Syllogeus* (1977).

Eddy, J. A. "The Maunder Minimum." In *Science* 192 (1976):1189.

Peccei, A. *Quale futuro?* Milan (1974).

MIT-Club of Rome. *I limiti dello sviluppo.* Milan (1972).

Vacca, R. *Il medioevo prossimo venturo.* Milan (1971).

THE END OF THE SOLAR SYSTEM
Books on cosmogony and stellar evolution, in general.

Asimov, I. *Catastrophe by Choice.* New York (1980). [This also deals with arguments that will be presented and discussed in Chapter 9.]

CHAPTER 6   LIFE ELSEWHERE

Sagan, C. *Cosmo.* Milan (1981).

Encranaz, P. *Materia e vita negli spazi interstellari.* Rome (1980; updated edition of the 1974 French original).

Shklovskij, J. S., and Sagan, C. *La vita intelligente nell'universo.* Milan (1980; translation and updating of the 1966 original edition by Libero Sosio).

Hoyle, F., and Wickramasinghe, C. *La nuvola della vita.* Milan (1979).

Morrison, P., Billingham, J., and Wolfe, J. *The Search for Extraterrestrial Intelligence.* NASA SP-419 (1977).

Cavaliere, A. "La comunicazione extraterrestre." In *La riscoperta del cielo.* Milan (1976).

Birand, F., and Ribes, J. C. *Le civiltà extraterrestri.* Milan (1971).

CHAPTER 7   TOWARD THE ORIGIN OF EVERYTHING

Herbst, W., and Assousa, G. E. "Supernove e formazione di stelle." In *Le Scienze* (October 1979).

Marsakov, V. A., and Suchov, A. A. "Orbital Elements and Kinematics of Halo Stars and the Hold Disk Population; Evidence for Active Phases in the Evolution of the Galaxy." In *Soviet Astronomy* 22 (1978):270.

Weliachew, L. "La dinamica delle galassie a spirale." In J. Lequeux, *L'astrofisica.* Rome (1978).

Demarque, P., and McClure, R. D. "Stellar Population of the Disk and Halo of the Galaxy." In *The Evolution of Galaxies and Stellar Populations.* New Haven (1977).

Tinsley, B. M. "Connection between Chemical and Dynamical Evolution." In *Chemical and Dynamical Evolution in Our Galaxy.* IAU Colloquium 45. Torun (1977).

Cowley, Ch. R. "Nuclear and Differentiated Abundance Patterns." In *Sky and Telescope* 52 (1976):236, 341.

Castellani, V. "La struttura e l'evoluzione della Galassia." In *Giornale di Astronomia* 1 (1975):189.

Gratton, L. "Popolazioni stellari ed evoluzione galattica." In *Giornale di Astronomia* 1 (1975):13.

Gratton, L. *Relatività, cosmologia astrofisica.* Turin (1968).

Eggen, O. J., Lynden-Bell, D., and Sandage, A. R. "Evidence from the Motions of Old Stars That the Galaxy Collapsed." In *Astrophys. J.* 136 (1962):748.

CHAPTER 8   FROM THE BIG BANG TO THE GALAXIES

Structure and Evolution of Normal Galaxies. Cambridge (1981). [For a clear synthesis of the arguments discussed in this book see also McCrea, W. H. "Origin and Early Evolution of Galaxies." In *Vistas in Astronomy* 23 (1979):173.]

Bertola, F. "What Shapes Are Elliptical Galaxies?" In *Sky and Telescope* (May 1981).

Davies, P. C. W. *Spazio e tempo nell'universo moderno.* Bari (1980).

Meier, D. L., and Sunyaev, R. A. "Galassie primordiali." In *Le Scienze* (January 1980).

Hack, M. "Il Nobel a Penzias e Wilson per la scoperta della radiazione cosmica a 3 K." In *Le Scienze* (October 1978).

Kleczek, J. *The Universe.* Dordrecht (1976).

Sciama, D. W. *Cosmologia moderna.* Milan (1973).

Gamow, G. *La creazione dell'universo.* Milan (1956).

CHAPTER 9   WHICH UNIVERSE?

Barrow, J. D., and Silk, J. "La struttura del primo universo." In *Le Scienze* (June 1980).

Masani, A. *Storia della cosmologia.* Rome (1980).

Dyson, F. J. "Time without End: Physics and Biology in an Open Universe." In *Rev. Mod. Phys.* 51 (1979):447.

Islam, J. N. "The Long-Term Future of the Universe." In *Vistas in Astronomy* 23 (1979):265.

Abell, G. O. "Cosmology. The Origin and Evolution of the Universe." In *Mercury* 7 (1978):45.

Gott, J. R., Gunn, J. E., Schramm, D. N., and Tinsley, B. M. "Cesserà l'espansione dell'universo?" In *Le Scienze* (July 1976).

CHAPTER 10   FROM THE PAST TO THE FUTURE

Davies, P. C. W. *Universi possibili.* Milan (1981).

Rees, M. J. "Our Universe and Others." In *Q. Il R. Astr. Soc.* 22 (1981):109.

Carr, B. J., and Rees, M. J. "The Anthropic Principle and the Structure of the Physical World." In *Nature* 278 (1979):605.

Davies, P. C. W. *Spazio e tempo nell'universo moderno.* Bari (1979).

Layzer, D. "La freccia del tempo." In *Le Scienze* (April 1976).

Layzer, D. "The Arrow of Time." In *Astrophys. J.* 206 (1976):559.

Davies, P. C. W. "How Special Is the Universe?" In *Nature* 249 (1974):208.

Collins, C. B., and Hawking, S. W. "Why Is the Universe Isotropic?" In *Astrophys. J.* 180 (1973):317.

Dyson, F. J. "L'energia nell'universo." In *Le Scienze* (December 1971).

INDEX